Logic in Elementary Mathematics

Robert M. Exner
and
Myron F. Rosskopf

Dover Publications, Inc.
Mineola, New York

Bibliographical Note

This Dover edition, first published in 2011, is an unabridged republication of the work originally published in 1959 by the McGraw-Hill Book Company, Inc., New York.

Library of Congress Cataloging-in-Publication Data

Exner, Robert M.
Logic in elementary mathematics / Robert M. Exner and Myron F. Rosskopf
p. cm.
Originally published: New York : McGraw-Hill, 1959.-
Summary: "This accessible, applications-related introductory treatment explores some of the structure of modern symbolic logic useful in the exposition of elementary mathematics. Topics include axiomatic structure and the relation of theory to interpretation. No prior training in logic is necessary and numerous examples and exercises aid in the mastery of the language of logic. 1959 edition"—Provided by publisher.
Includes index.
ISBN-13: 978-0-486-48221-7 (pbk.)
ISBN-10: 0-486-48221-9 (pbk.)
1. Mathematics—Philosophy. 2. Logic, Symbolic and mathematical. I. Rosskopf, Myron Frederick. II. Title.

QA9.E96 2011
510.1—dc22

2010052658

Manufactured in the United States by Courier Corporation
48221901
www.doverpublications.com

PREFACE

The threefold purpose of the present text is:

1. To present some of the structure of modern symbolic logic that is useful in the exposition of elementary mathematics.

2. To use the logic in discussion of formal aspects of elementary mathematics.

3. To present the material in such a way that it will be accessible to those whose mathematical training includes the standard elementary courses, but who may lack any formal courses in logic.

It is not our primary interest to develop the logic for its own sake, but rather as a language for discoursing about mathematics. We do not attempt an abstract development of logic from axioms. To justify the logical principles assumed, we lean heavily on the intended interpretation of the logic in mathematics. All mathematical interpretations are elementary in nature. Exercises and examples are included in appropriate places as an aid to mastery of the logical language, and those that are mathematical are drawn from elementary algebra and geometry and trigonometry.

We limit ourselves generally to certain formal and structural aspects of mathematics. It is true that in the first chapter, to highlight our aims by contrast, we do talk about problem solving, mathematical intuition, and the creative side of mathematics. We shall have little further to say, however, about these important matters in succeeding sections, but shall have much to say about the axiomatic structure of mathematics, about *proof* or *demonstration*, about the relation of *theory* to *interpretation*, about *variables*, *equality*, and *functions*, and other such notions of a formal nature. The notions treated are just those that, in our experience, are most often treated fuzzily in traditional exposition. Indeed, some of the notions we shall treat are entirely tacit in traditional exposition.

The authors owe much to their colleagues and students. The exposition, including exercises, was tested upon both. We owe a greater debt to the mathematicians and logicians whose results we use, and whose clarity of exposition we hope to have approximated. Some of those whose writings have been most helpful are W. Ackermann, Alonzo Church, Irving M. Copi, D. Hilbert, Stephen C. Kleene, Hugues Le Blanc, Willard Van Orman Quine, J. Barkley Rosser, and Alfred Tarski.

Robert M. Exner
Associate Professor of Mathematics
Syracuse University

Myron F. Rosskopf
Professor of Mathematics
Teachers College
Columbia University

CONTENTS

GLOSSARY OF SPECIAL SYMBOLS AND ABBREVIATIONS

Symbol	Page
“ ”	10
$\sim$	20
&	22
$\vee$	24
$\rightarrow$	27
$\leftrightarrow$	33
$\frac{P,\ Q}{PQ}$, etc.	50
mod pon, m.p.	50
conj inf	51
simp; conj simp	51
hyp syll	52
contrapos inf	52
$\vdash$	79
ε	81
ded prin	89
$H(\quad)$, $F(\quad)$, etc.	127
(x); $\forall(x)$	132
$(\exists x)$	135
IU	143
PGU*	146
IE	152
PGE	153
PGU	166
R	174
IS	174
$F^*(y)$	174
ISC	176
e	183
$\circ$	184
x^{-1}	188
1-1	192
$\cong$	193
F [mod p]	200
disch assump	223 (see 178, 180)
$\otimes$	224
$\lvert a \rvert$	228
A_n, $\{A_n\}$	234
contra	244
subst VSF	244
quant conv	247 (see 138)

The number opposite each symbol indicates the page on which it is explained.

I: MATHEMATICS, FORMAL LOGIC, AND NAMES

1.1 MATHEMATICS AND MATHEMATICIANS

That mathematics is one of the oldest and richest of the disciplines is generally conceded. What is not so commonly understood is that mathematics is also one of the most rapidly expanding of the disciplines. Several times in the life of a mathematician he can expect to be asked, "But what do you do? Is there anything new to do in mathematics?" Scientists and engineers are less subject to this sort of question, for everyone knows that these men are plunging ahead to marvelous new discoveries in their fields. But there is a widespread feeling that mathematics is a precise but static body of knowledge that was pretty well roughed out by the Greeks and other early peoples, leaving little for modern man to do but polish and refine.

Nevertheless, a mathematician finds it much easier to describe what he does than to answer the question "What is mathematics?" Attempts to define mathematics range from pithy sentences to books in multiple volumes. It has been variously described as a language, a method, a state of mind, a body of abstract relations, and so on. A thoroughly satisfactory definition is not easily attained.

In describing what he does, a mathematician is likely to list creation of new mathematics as one of his important activities. Creation of new mathematics involves first discovering new mathematical relationships and then proving them. In searching for new mathematics, the mathematician does not limit his methods. He uses his imagination and ingenuity to the utmost. He experiments and he guesses. He uses analogies. He may build machines and physical models. He makes some use of deductive methods but much more use of inductive methods. In the heat of the search nothing is illegal.

Once he has guessed a new relationship and has satisfied his intuitions that the new idea is reasonable, the mathematician has what he calls a *conjecture*. He next tries to change the conjecture into a theorem by proving (or disproving) that the conjecture follows logically from the assumptions and proved theorems of known mathematics. In constructing a proof, he sharply limits himself to the methods of deductive logic. He does not allow his proof to depend on experiment, intuition, or analogy. He tries to present an argument satisfying a set of rules that mathematicians have agreed such arguments must satisfy.

A very famous conjecture of mathematics is due to Goldbach (1690–1764), who conjectured that every even integer greater than 4 is the sum of two odd prime numbers.[1] We do not know how Goldbach was led to this discovery, or just how he satisfied himself as to its reasonableness, but a reasonable way to test the conjecture would be to test some particular cases:

$$
\begin{aligned}
6 &= 3 + 3 \\
8 &= 5 + 3 \\
10 &= 5 + 5 = 7 + 3 \\
12 &= 7 + 5 \\
14 &= 7 + 7 = 11 + 3 \\
16 &= 11 + 5 = 13 + 3 \\
18 &= 11 + 7 = 13 + 5 \\
20 &= 13 + 7 = 17 + 3
\end{aligned}
$$

The first few cases work out nicely. To increase one's confidence in the conjecture, one might try some larger numbers chosen at random:

$$
\begin{aligned}
90 &= 83 + 7 \\
100 &= 97 + 3 \\
108 &= 97 + 11 \\
1286 &= 1279 + 7
\end{aligned}
$$

[1] A prime number is an integer that has no positive integral factors other than itself and one. For various reasons, it is best to exclude 1 from the set of primes, although it meets the requirements of the definition.

The more special cases one verifies, the more confidence one has in the conjecture, but of course no amount of verified cases constitutes a proof. No one has published a proof of this conjecture to date; neither has anyone found an even number greater than 4 and shown that it is not the sum of two odd primes. Thus the truth of Goldbach's conjecture remains an open question.

Another well-known conjecture concerns the sequence:

$$3, 5, 17, 257, 65537, 4294967297, \ldots, 2^{2^n} + 1, \ldots.$$

The mathematician Fermat [1601(?)–1665] made the conjecture that all the terms in this sequence are prime numbers. It is not too hard to verify the conjecture for the first five terms of the sequence, corresponding to $n = 0, 1, 2, 3, 4$. To verify that 257 is prime, for instance, it is only necessary to show that it is not divisible by any of

$$2, 3, 5, 7, 11, 13.$$

One can determine if any integer N is prime by testing to see if it is divisible by any lesser prime. In fact, one easily sees that it is sufficient to test for divisibility by the primes that are not greater than the square root of N. In the case of the fifth term 65537, one would then test the primes less than 256. Since there are 54 of these, the task will be tedious. To determine, by this method, whether the sixth term 4294967297 is prime would require testing each of the primes less than 65536 to see if any one of them divides 4294967297 exactly. Since there are 6543 primes less than 65536, one would not look forward to the task. In fact, mathematicians will not engage in such computations as this if they can possibly avoid them, particularly since it is clear that the computations become much worse for later terms of the sequence.

There is evidence that Fermat felt quite strongly that his conjecture was reasonable; yet the mathematician Euler (1707–1783) found that the huge number can be factored as follows:

$$4294967297 = (641)(6700417).$$

This remarkable discovery settled Fermat's conjecture once and for all. The terms of the sequence are not all primes. Further study of this sequence has led to other conjectures and theorems of considerable interest, but we cannot pursue the matter here.

Having seen an example of an open conjecture and of a disproved conjecture, one naturally looks for the discussion to be rounded out with an example of a proved conjecture. But almost any theorem of mathematics is an example, for nearly all theorems were conjectured before they were proved. It is generally a reasonable conjecture that provides one with a motive, and the courage, to attempt the arduous task of forging a proof. Imagine being asked to prove a relationship between the hypotenuse and legs of right triangles if you have never heard of the Theorem of Pythagoras!

The mathematician usually presents his results as a series of theorems with proofs. This method of presentation tends to obscure the thought processes which led to his conjectures, as well as the methods he used to discover his proofs, but it has the advantages of precision and economy of exposition. The requirement that new additions to mathematics be proved exerts a profound unifying effect on the development of the discipline. It ensures that the body of mathematics grows in an orderly way, and that new mathematics is immediately related to existing mathematics. The solid organization of the structure of mathematics makes it possible to leap ahead oftener and further than is possible in some of the new disciplines that do not yet have such well organized structures.

The system of deductive logic provides mathematics with

a. An efficient means of exposition, and

b. An efficient means of organizing its subject matter.[1]

In what follows, we shall be concerned with this formal, logical side of mathematics. The logic provides a symbolic language with which to talk about mathematics with ease and precision. We shall also have to introduce a symbolic language in order to talk about the logic with ease and precision.

We shall have little to say in this book about the creative, conjectural side of mathematics, but this should not be construed as indicating any judgment about the relative importance of the formal side and the creative side of mathematics. Both sides are essential to the discipline.

1.2 FORM AND MEANING

Consider the case of a mathematician who has arrived at a reasonable conjecture. However plausible or seductive the conjecture, he must view it with suspicion until he knows whether it is "true". We use quotation marks here because in this case "truth" means consistency with the body of accepted mathematical statements. To verify this consistency, the mathematician tries to construct a *proof* that will show that the statement of the conjecture *follows logically* from accepted mathematical statements. If he succeeds in this endeavor he considers that he has proved his conjecture, and henceforth has the same attitude toward it as toward any previously accepted mathematical statement. Another mathematician will accept the new statement only if he agrees that the proof is correct, or if he can construct a proof of his own.

It appears that there must be some general agreement among mathe-

[1] Of course, logic also provides these means for the sciences and other disciplines which use them in varying degrees and with varying success.

maticians as to the meaning of the phrase "follows logically" as used above, and some generally accepted rules for constructing proofs. We shall mean by *formal logic* just such a system of rules and procedures used to decide whether or not a statement follows from some given set of statements.

Suppose a mathematician has arrived at a conjecture D through introspection, intuition, and verification of particular instances. He feels sure that it is correct. Now he is faced with the necessity of establishing D according to the strict rules of deduction. The mathematician might reason, "I know that A was proved to my satisfaction by Xoxolotsky in 1951. Clearly B follows from A. D is my conjecture and it follows from C. Aha! B implies C, so I know that D follows from A and my conjecture is proved."

Formal logic will not indicate what sequence of statements to choose for a proof, but once such a sequence of statements is chosen, the logic is designed to decide whether B follows from A, C from B, D from C, and ultimately whether D follows from A. Furthermore, the logic is designed to decide such questions purely on the basis of the forms of the statements involved, without reference to their meanings.

The point of view expressed in the preceding paragraph can be illustrated by familiar examples from Aristotelian logic. Sooner or later in books on logic it is traditional to present an argument about Socrates:

1. All men are mortal.
2. Socrates is a man.

∴ 3. Socrates is mortal.

The Socrates Argument is regarded as an example of a valid form of reasoning in Aristotelian logic; that is, according to the rules of this logic, if any three statements have the forms (1), (2), (3), then (3) follows from (1) and (2). The argument is correct no matter what meanings statements (1), (2), and (3) have; all that is required is that they have the forms of (1), (2), and (3). The three statements in the example would generally be considered true (or perhaps more accurately, at a suitable time in history they were considered true). In any event, to say about the statement "All men are mortal" that it is true is to voice a judgment based on its meaning, and meaning is not in question when we are considering the validity of an argument.

Consider another example of the foregoing valid form of argument:

1. All people who like classical music are baseball players.
2. Babe Ruth liked classical music.

∴ 3. Babe Ruth was a baseball player.

Most people would judge (1) to be false, would agree that (3) is true, and would not feel qualified to judge the truth value of (2). Such judgments about the meanings of the statements do not affect the validity of the argument. If statements (1) and (2) are granted, then we are forced to grant the conclusion (3).

It is easy enough to construct a correct argument in this form whose conclusion is false. For instance:

1. All who ride by airplane were born after A.D. 1800.
2. Socrates rode by airplane.

$\therefore$ 3. Socrates was born after A.D. 1800.

The conclusion is a false statement. Yet, although the conclusion is ridiculous, the argument that led to it is valid.

The assertion "Truth is not the concern of logic" is too broad, since we shall want to say of arguments in the above form that if the first two statements are each true, then so is the third statement true. But in this sense, we arbitrarily assign the value "true" to each of the first two statements, and state as a rule of the logic that the third statement must then be assigned the value "true". The truth of the statements in terms of their meanings is not in question. Of course, we construct a system of logic in order to help us reason with meaningful statements. While it is the goal to develop methods for determining validity that are mechanical and based on form alone, still the valid forms should be such that from hypotheses that are true in the sense of meaning we obtain inescapable conclusions that are true in the sense of meaning.

The form of the Socrates Argument is not determined by the words "men", "mortal", "Socrates" but by the other words in the statements and by the order in which the statements occur. If the words "men", "mortal", "Socrates" are replaced, respectively, by the terms "person who likes classical music", "baseball player", "Babe Ruth", then we have an argument which is still in the form of the Socrates Argument but whose meaning is the same as that of the second example. (We overlook here some grammatical discrepancies of tense and number.) Since the form of the argument is not dependent on the particular terms used, we are tempted to exhibit the form as

1. All _____ are _____.
2. _____ is _____.

$\therefore$ 3. _____ is _____.

However, this pattern is not adequate since it seems to permit six different terms for distribution in the blank spaces, and the two examples deal with only three terms each. We are then led to exhibit the form as

1. All $\underline{\quad M \quad}$ are $\underline{\quad P \quad}$.
2. $\underline{\quad S \quad}$ is $\underline{\quad M \quad}$.

∴ 3. $\underline{\quad S \quad}$ is $\underline{\quad P \quad}$.

By this scheme of lettering the blanks, our intent is to indicate that whatever the three terms chosen to fill in the blank spaces, those blank spaces designated by the same letter must be filled in with the same term. It will be simpler to indicate a blank space by a letter without any underscoring, thus

1. All M is P.
2. S is M.

∴ 3. S is P.

In Aristotelian logic, an important form of argument called the "syllogism" has forms somewhat similar to that of the Socrates Argument. The statements appearing in syllogistic arguments are simple declarative sentences, usually called propositions, that are classified under four headings:

Classification	*Examples*
A: Universal and affirmative	All men are mortal.
E: Universal and negative	No men are mortal.
I: Particular and affirmative	Some men are mortal.
O: Particular and negative	Some men are not mortal.

The logical form of a proposition is:

Subject—a form of the verb *to be*—*predicate.*

The terms in the subject and predicate can usually be satisfactorily interpreted as names of classes, and the relation between subject and predicate is taken to express the inclusion of all, part, or none of one class in the other. An example of a syllogism is:

1. All monkeys are tree climbers.
2. All marmosets are monkeys.

∴ 3. All marmosets are tree climbers.

A syllogism is an argument consisting of two propositions called *premises* and a third proposition called the *conclusion.* In the premises two terms are each compared with a third, and as a result there is a relationship between the two terms that is expressed in the conclusion. Traditionally, the term in the subject of the conclusion is called "S", and that in the predicate "P"; the term to which they are both compared is called the *middle term,* and is denoted by "M". With this notation, the form of the foregoing syllogism can be expressed:

1. All M is P.
2. All S is M.

∴ 3. All S is P.

This syllogistic form is said to be in the first *figure*, and its *mood* is "AAA". The figure is determined by the position of the middle term, and the mood by the kinds of propositions and the order of their occurrence. Since interchanging the order of the premises does not change the relationships they express, it does not matter which is written first. It is traditional to write the premise comparing M and P first. Subject to this tradition, there are just four possible distinct figures:

I	II	III	IV
MP	PM	MP	PM
SM	SM	MS	MS
SP	SP	SP	SP

Since each of the three propositions in a syllogism can have any one of the forms A, E, I, O, each figure can have 4^3 different moods. Hence, in all there are 256 different forms possible. Most of these forms are not considered to be correct forms of reasoning. In a traditional treatment, a set of rules is given for differentiating between the valid and invalid forms. Four of these rules are:

1. A syllogism having both premises negative is invalid.
2. In a valid syllogism, if one premise is negative, the conclusion is negative.
3. In a valid syllogism, if both premises are affirmative, the conclusion is affirmative.
4. A syllogism having both premises particular is invalid.

There are other such rules. The complete set of rules applied to the 256 possible forms of a syllogism identify 19 valid forms. According to these rules, the remaining 237 forms are classed as invalid.

This very brief excursion into Aristotelian logic serves to illustrate that methods of describing validity are formal and arbitrary. The rules for differentiating between valid and invalid syllogisms can be thought of as axioms of Aristotelian logic, and they are concerned solely with the forms of syllogisms, and not at all with meanings. All the words:

Syllogism	Negative
Figure	Premise
Mood	Conclusion
Affirmative	etc.

refer to various aspects of form, and are not concerned with any meanings that may be attached to the propositions in a syllogism. Of course, it is no accident that when we do consider the meanings of the propositions of a VALID syllogism, we find that if the premises are true then so is the conclusion true. For, however formal the system, it was created with just such an interpretation in mind.

In succeeding sections we want to describe a formal logical system suitable for the exposition and organization of modern elementary mathematics. We shall need a precise language for describing this system. It will be a language suitable for describing form, and it will be largely symbolic.

1.3 USE AND MENTION

Any book is sure to be full of statements. In this book, statements not only occur but are talked about and relationships between statements are discussed. In this sort of discourse it is well to be careful about the use of names. Names are *used* in statements to *mention* objects. Our language is so constructed that a statement about an object never contains that object but must contain a name of that object. Consider the statement

(1.1) This pencil has soft lead.

The expression "this pencil" names an object, and it is this name that is contained in the statement, rather than the object. If one were to take the pencil mentioned in (1.1) and hold up next to it a slip of paper with the words "has soft lead" written on it, he would probably succeed in conveying the sense of (1.1), but he would not be communicating by means of a sentence or statement.

No one dealing with a statement about a physical object is likely to confuse the object with its name. This kind of confusion is more likely to arise when the object mentioned in a statement is itself a name. In the statement

(1.2) Seattle is in the state of Washington,

the name of the city is used to mention it; there is little likelihood of confusing the name (a word) with the object (a city). Now it is reported that the name of the city mentioned in (1.2) is derived from the name of a Duwamish Indian. Suppose we attempt to convey this information by writing

(1.3) Seattle is derived from the name of a Duwamish Indian.

Since it makes no sense to say that a city (a large complex physical object) is derived from the name of an Indian, we conclude that the first word in

(1.3) cannot refer to a city, and that (1.3) is supposed to assert something about the *name* of some city.

The modes of expression in (1.2) and (1.3) are not compatible. The trouble is with (1.3), which violates the conventions regarding *use* and *mention* by employing as its first word the object about which (1.3) is supposed to be saying something. If (1.3) is considered to assert something about a city, it is a correctly formed sentence, but false. If (1.3) is considered to assert something about the name of a city, then it is not a sentence for the same reason that a pencil followed by the phrase "has soft lead" is not a sentence. The conventional way to deal with this problem of written exposition is to write

(1.4) "Seattle" is derived from the name of a Duwamish Indian.

The quotation marks are used to indicate that it is the name of the city that is the object of discourse, and not the city itself. For further illustration compare the statements

(1.5) Syracuse is in New York State.
(1.6) "Syracuse" has three syllables.
(1.7) "Syracuse" designates a city on Onondaga Lake.
(1.8) Syracuse is north of Ithaca.

It would be quite wrong to omit the quotation marks in (1.6) and (1.7) since a city cannot have three syllables, nor can it designate anything.

The combination of quotation marks with a word (or words) in their interior is called a *quotation*. A quotation names, or denotes, its interior, which is always an expression rather than a physical object. Thus it is correct to say

(1.9) "Syracuse" is used in (1.5) to mention Syracuse.

The name-forming device of quotation keeps us straight about what objects are mentioned in (1.9). Since the first expression in (1.9) is a quotation, we know that it is the name of the city that is being mentioned, whereas the last expression in (1.9) clearly mentions the city itself.

In each of (1.2), (1.5), and (1.8) the first expression is a name of a city. In each of (1.6), (1.7), and (1.9) the first expression is a name of a name of a city. Of course, we can carry this further by writing

(1.10) "'Syracuse'" is used in (1.6) to mention "Syracuse".

The first expression in (1.10) is a name of a name of a name of a city. We shall not need to go this far in the succeeding sections. We shall have to *mention* names, so will *use* names of names, but we shall not have to *mention* any names of names, so will not *use* any names of names of names.

In mathematics and logic, it will not do to confuse objects with their names. In logical discourse one can expect the name-forming device of quotation to be used freely. Mathematical writing is conventionally less careful in this matter, and one often has to determine from context what the objects of discourse are. Generally the context is clear and no confusion of object and name is likely; yet there are some areas of mathematics in which this kind of confusion has occurred and has given rise to furious controversy. One such controversy arises periodically concerning the distinction between zero as a number and as a placeholder. Those who maintain that zero is a number like to cite contexts for zero, such as

$$0 + 3 = 3,$$
$$0 \cdot 4 = 0,$$
$$5 - 5 = 0,$$

to support the notion that zero behaves like other numbers. Supporters of zero as a placeholder like to cite its use in expressing such numbers as 20, 1000, 207, etc. The controversy arises from a confusion of name and object. While one side talks about 0, the other side talks about "0", and neither is aware that they are talking about different things. The controversy disappears when careful distinction is made between 0 and "0". Consider the statement

(1.11) The sum of 0 and 3 is 3.

This statement mentions the numbers 0 and 3 by using the symbols "0" and "3". Now consider the statement

(1.12) 20 can be expressed by "2" followed by "0".

This statement describes one way that 20 can be expressed by symbols. The symbols are "2" and "0", which are names of the numbers 2 and 0, respectively. The statement (1.11) mentions the number 0, and (1.12) mentions the name "0".

The distinction between 0 and "0", or between 7 and "7", is just the distinction between a number and a numeral. Numerals are special names for numbers. A given number may be mentioned by using various different numerals. For instance, in the statement

(1.13) 7 is greater than 1,

arabic numerals "7" and "1" are used to mention the numbers 7 and 1. One could as well use the Roman numerals "VII" and "I" to mention 7 and 1; thus

(1.14) VII is greater than I,

or use Greek numerals to write

(1.15) ς is greater than α.

One might say that the number 7 has no nationality, but the numerals "7", "VII", and "ς" do have nationalities. Statements (1.13), (1.14), and (1.15) are undeniably different, but they both mention the same objects in the same way and have the same meaning.

It is a general principle that one may substitute for a name of some object in a given statement any other name of the same object without altering the meaning of the given statement. It is easy to misuse this principle when applying it to a statement in which objects and names are confused. Consider the following short argument:

(1.16)
1. The denominator of 14/18 is even.
2. 7/9 is another name for 14/18.

∴ 3. The denominator of 7/9 is even.

At first glance, the argument is apparently correct and the first two statements apparently true, but the conclusion seems false, so we seem to have a paradox.

Let us examine each statement in the argument (1.16) to ensure that in each the intended objects are correctly mentioned. In (1) the object apparently mentioned is the denominator of 14/18, but there is no such object; 14/18 is a number and cannot have a numerator or a fraction bar or a denominator. The words "numerator", "fraction bar", "denominator" designate parts of symbols used to denote rational numbers. Thus the object that should be mentioned in (1) is "14/18". Similar remarks apply to statement (3). As for (2), the first expression in it is clearly intended to mention a name of a number, and not a number, so quotation must be used.

Properly rewritten (1.16) becomes

(1.17)
1′. The denominator of "14/18" is even.
2′. "7/9" is another name for 14/18.

∴ 3′. The denominator of "7/9" is even.

Now it begins to become clear why (2′) does not justify the substitution in (1′) that yields (3′). The object mentioned in (1′) is the name "14/18". To make the substitutions, we would need to know some other name for "14/18", but (2′) just gives us another name for the number 14/18, so (2′) is of no use here. On comparing (1′) and (3′) we see that neither could be obtained from the other by substitution because the objects *mentioned* are different; hence the names used to mention them are not interchangeable.

The same sort of fallacy is illustrated by the nonmathematical argument

1. "Mark Twain" is a pseudonym.
2. "Samuel Clemens" is another name for Mark Twain.

∴ 3. "Samuel Clemens" is a pseudonym.

The foregoing examples contain several cases of different words or symbols that are names of the same object. Some of these cases are listed below:

Name	*Type of object named*
the first expression in (1.5), "Syracuse"	a city
the name of the city mentioned in (1.2), "Seattle"	a city
the first expression in (1.6), "'Syracuse'"	a name
"7", "VII", "ζ"	a natural number
"7/9", "14/18"	a rational number
"Samuel Clemens", "Mark Twain"	a person

A conventional way of expressing

"7" and "VII" are names of the same object

is to write

$$7 = \text{VII}.$$

Similarly,

$$7/9 = 14/18$$

means

"7/9" and "14/18" are names of the same object.

In the same sense we may write

Mark Twain = Samuel Clemens.

We shall take this interpretation of "=" in all subsequent exposition. We shall also use the substitution rule that in any statement, one name may be substituted for another as long as both are names of the same object. This interpretation of equality, and the substitution rule, are consistent with mathematical practice.

1.4 STATEMENTS ABOUT STATEMENTS

The conventions regarding use and mention apply when we want to write about statements. In order to mention a given statement we have to use a name of it. The device of quotation is a convenient way of forming the required name. If we want to say of statement (1.5) that it is true, we write

(1.18) "Syracuse is in New York State" is true.

Other examples of statements about statements are

(1.19) In "Syracuse is in New York State" both a city and a state are mentioned.
(1.20) "All women are faithless" is unbelievable.
(1.21) "All cats are black" and "No cat is black" are both false.
(1.22) "'Syracuse is in New York State' is true" is believable.
(1.23) "He lies" is blunter than "He prevaricates".

A way of writing about a statement without using quotation marks is to display the statement centered on the page and on a new line. We have used this device repeatedly; indeed we have just used it six times in this section. Instead of (1.18) we can write

(1.24) Syracuse is in New York State

is true. To center and display material in this fashion is tantamount to quoting it, and permits some reduction in the number of quotation marks needed.

Yet another way to avoid quotation marks is to use the numeral at the left-hand margin as a name for the material displayed on a line with it. Thus (1.18) can now be written

(1.24) is true.

By the same device, the complicated statement (1.22) can be written

"(1.24) is true" is believable;

or even,

(1.18) is believable.

Finally, we shall never use a letter as the name of a statement. When we write

(1.25) A: John is dead,

we intend "A" as a symbolic *translation* of "John is dead", just as "Jean est mort" is a French translation of it and "Johann ist tot" is a German translation of it. With this agreement, we can write about the statement (1.25), or the statement "A", or the statement "John is dead", and have it understood that we are simply mentioning a single statement in three different ways.

We have dwelt on these devices at some length because they will all be used repeatedly in the subsequent exposition to ensure that it is always clear about what we are talking. It must be confessed that very often omission of quotation marks causes no confusion, and in a good deal of mathematical exposition they are not used. We shall sometimes omit quotation marks in what follows but will try to avoid this practice whenever real confusion might result.

EXERCISES

1. Put quotation marks where necessary in each of the following sentences in order to make them meaningful statements.

a. Chicago has three syllables.
b. The sum of 16 and 0 is 16.
c. In the preceding statement 16 and 0 are used to mention the numbers 16 and 0, respectively.
d. Another way to express 20 is by an X followed by another X.
e. The numerator of 8/13 is even.
f. 2/3 is another name for 6/9.
g. It is 6/9 that has a numerator, fraction bar, and denominator, but 6/9 has none of these.
h. The denominator of 5/9 is odd.
10/18 is another name for 5/9.
∴ The denominator of 10/18 is odd.

2. Indicate the kind of object named by each of the following symbols.

a. F.D.R. *Ans.* The object named is a former Democratic president.
b. 3 *c.* "5"
d. A.D. *e.* △ABC
f. Russia *g.* "Russia"

3. Write a sentence in which you use correctly:

a. H-bomb *b.* "H-bomb"

4. Mention Illinois in a sentence.
5. Use "Illinois" in a sentence.
6. Can you use Illinois in a sentence?
7. Mention a name of New York State in a sentence.
8. Write a true statement about all college students.
9. Write a true statement about "all college students".
10. Write a true statement about "To be or not to be is the question".
11. Rewrite those of the following statements that you think need quotation marks. If necessary, support your answer with reasons.

a. Bookkeeper is one of the few words in the English language containing a double k.
b. If 18 is divided by 3, the result is 6.
c. Big Ben has a pleasant sound.
d. Big Ben is alliterative.

e. Angle ABC indicates that B is the vertex of an angle.
f. We use 3, 1, and 5 to write 315.
g. One way to name 47 is to write 4 followed by 7.
h. 5 is used in the direction on page 5 you will find . . . .
i. ABCD or q can be used as names for a quadrilateral.

12. Restate the meaning of "$x = y$" in your own words.

13. "$\triangle$ABC = $\triangle$DEF" should mean that "$\triangle$ABC" and "$\triangle$DEF" are different names for the same triangle. Discuss various interpretations of "$\triangle$ABC = $\triangle$DEF" as used in plane geometry.

14. If A and B are points, then "AB" denotes the line segment determined by A and B. This is the customary notation of plane geometry. Yet "AB = CD" is not interpreted as a statement about a single line segment, but as a statement about two different line segments. Discuss this point.

II: THE STATEMENT CALCULUS

2.1 STATEMENTS AND STATEMENT FORMS

If the notion is held that validity is a matter of form rather than a matter of meaning, then in investigating questions of validity it is reasonable to use a symbolism that indicates form alone. With a suitable symbolic language we can hope to lay bare the formal structure of argument without being distracted by particular meanings.

For example, let "P" be a translation of the statement

P: All men are mortals.

"P" can be considered a statement of relationship between two classes; that is, the class of *men* and the class of *mortals*. The words "all" and "are" indicate the relation, while the words "men" and "mortals" are names for the classes so related.

Consider now the expression

P^*: All x are y.

Unless "x" and "y" are names of known classes, the expression "P^*" is in a sense meaningless. For instance, it makes no sense to assert that "P^*" is true, or that it is false. However, if "x" and "y" in "P^*" are taken to be placeholders for class names, then "P^*" can serve to indicate a type of *statement form*. The statement "P" has this form. The first statement of the second argument on page 5 also has the form of "P^*". In fact, you can obtain it from "P^*" by replacing "x" by the name "people who like classical music" and "y" by the name "baseball players".

An expression such as "P^*" that employs symbols "x" and "y" as placeholders for class names is called a *propositional function* with variables x and y. Such propositional functions will be considered in Chap. V. For the present, we shall restrict ourselves to the simpler question of relationships between statements considered as units. Capitals "P", "Q", "R", . . . will be used as translations of complete statements. No attempt will be made in the notation to indicate the internal structure of a statement unless it is a compound of simpler statements.

With this restriction we shall be unable to handle even such a simple argument as is represented by the Socrates Argument, page 5, since the validity of such an argument depends on subject and predicate relations that require the logic of propositional functions for analysis. Even so, we shall be able to express formally many arguments.

For the present, we take a statement in its simplest form to be a simple declarative sentence that is either true or false, but not both. For our purposes, the statement

Smith is intelligent

can be recast in the form

Smith is an intelligent man.

That is, the individual "Smith" is asserted to be a member of the class "intelligent men". In a similar fashion, the statement

All bats fly

can be recast in the form

All bats are animals that fly.

That is, it is asserted that the class "bats" is included in the class "animals that fly".

It is not easy to give criteria for determining when two statements assert the same proposition. The foregoing restatements may not express the exact shades of meaning of the original statements, but the difference is not serious, and the restated forms are easier to deal with formally. Such restatements work out well enough in formal analysis of mathematical statements. In what follows, we shall always consider the simple statements used to mean assertions of class inclusion even though they may not appear in that form.

Each of the following statements illustrates a different relationship between subject and predicate.

1. Washington was the first president of the United States.
2. Washington was wise.
3. Washington was crossing the Delaware.
4. Washington was a soldier.

In (1), the relationship expressed is that of identity; in (2), that of subject and attribute; in (3), that of agent and action; in (4), that of inclusion of an individual in a class. The statements in (2), (3), (4) are readily recast as statements of class inclusion. In (1), it is asserted that "Washington" and "the first president of the United States" are names of the same thing or object. We do not attempt to restate this in terms of class inclusion, and shall avoid statements of this type for the moment.

We have described the simplest statements as simple declarative sentences. We shall also wish to regard as statements, compounds of simple statements formed by using connective words such as "and", "or", "if _____ then_____". In what follows we shall try to develop a precise symbolic language for expressing such composition as a purely formal operation.

EXERCISES

List the individual and the class, for each of Examples 1 to 6. Put these under column headings of *Individual* and *Class*.

1. Eagles are animals that fly.
2. President Lincoln was an intelligent man.
3. Saturday is a day of the week.
4. 8/9 is a rational number.
5. Doctors are good reasoners.
6. The sum of 6 and -7 is an integer.

Recast each of Examples 7 to 14 as a statement of class inclusion. For example, "All bats fly" could be recast as "The class of bats is included in the class of animals that fly".

7. John is a high school student.
8. Triangle ABC is isosceles.
9. The number $\sqrt{3}/4$ is irrational.
10. A quadrilateral with two pairs of parallel sides is a parallelogram.
11. All real numbers are complex.
12. Any teacher is an educator.
13. Each state is in the Union.
14. Circles are conics.

2.2 NEGATION

The negation of a statement is formed by means of the word "not". If "P" is a translation of a statement, then the negation of the statement is translated "$\sim P$". "$\sim P$" is read "not-P"; the symbol "$\sim$" is called "curl" or "twiddle" or "tilde". (The notion of "$\sim P$" is that of asserting the falsity of "P". If, indeed, "P" is considered to be false, then "$\sim P$" will be considered to be true. If "P" is a translation of

New York is a city,

then "$\sim P$" is translated

Not, New York is a city.

In everyday language, the negative particle "not" is generally associated with the verb. Consequently, a restatement of "$\sim P$" in everyday language would read

New York is not a city.

In making such restatements, we want to be sure that the statement and its restatement assert the same proposition. For simple statements, the correct and natural restatement is easily discovered, but care must be taken with statements containing the quantifying words "all" and "some".

Consider the problem of restating in natural language the negation of

Q: All integers are real numbers.

Then, for the negation of "Q" we have

$\sim Q$: Not, all integers are real numbers.

To recast "$\sim Q$" in the form "All integers are not real numbers" is at the least ambiguous. The ambiguity becomes apparent upon reading the sentence, first stressing the word *all*, then stressing the word *not* The meaning in the first reading is expressed by the statement

V: Some integers are not real numbers.

For the second reading, the meaning is expressed by the statement

W: No integers are real numbers.

Since it is quite clear that "V" and "W" do not assert the same proposition, they cannot both be correct restatements of "$\sim Q$". According to our definition of negation, "$\sim Q$" must be opposite in its truth value to "Q". Any correct restatement of "$\sim Q$" must also have this property. The statement "W" fails this test, for while "Q" and "W" cannot possibly both be true, they can conceivably both be false. Indeed, both statements would be false if it were the case that there is exactly one integer which is not a real number. However, if "V" is tested in the same way, it is found to have the required properties, and may be taken as a correct restatement of "$\sim Q$".

The form "All ______ are not ______" is seen to be ambiguous. Its meaning depends upon vocal stress, or on context. Such a form should be avoided in mathematics and in logic.

As another example of the problem of restatement, consider

S: All angles can be trisected using straightedge and compass alone.

This statement is known to be false. Some students, knowing that "S" is false, jump to the conclusion that you can't trisect an angle. By this they mean

U: No angle can be trisected by using straightedge and compass alone.

But "U" is known to be false, and thus cannot be a correct restatement of "$\sim S$". Since both "S" and "U" are false, according to our notion of negation, "$\sim S$" and "$\sim U$" are true statements. As was seen in the previous example concerning integers and real numbers, a correct restatement of "$\sim S$" is

Some angles cannot be trisected using straightedge and compass alone.

Another common restatement of "$\sim S$" is

There exists at least one angle that cannot be trisected by using straightedge and compass alone.

It is a standard demonstration of higher algebra that an angle of 120° is such an angle.

"$\sim U$" is correctly translated

Not, no angle can be trisected by using straightedge and compass alone,

which is grammatically very awkward. It is better to use the restatement

Some angles can be trisected by using straightedge and compass alone;

or the form

There exists an angle that can be trisected by using straightedge and compass alone.

Angles of 90° and 180° are such angles.

EXERCISES

Write a grammatical restatement of the negation of each of the following sentences.

1. Chicago is the Windy City.
2. John is not at his home.
3. Solid geometry does not exist as a separate course.
4. This textbook has many exercises.
5. a $\parallel$ b
6. All rational numbers are real.
7. Some cats are black.
8. Some pairs of lines in a plane are parallel.
9. No imaginary numbers are real.
10. All people are intelligent.
11. No slow learners attend this school.
12. All courses overlap.
13. None of us may go.
14. Some of these references are not relevant to my subject.
15. Some triangles are not isosceles.
16. All brilliant persons are teachers.
17. No employees are honest.
18. Some unpleasant statements are not true.
19. All fractions with a common denominator are added by writing the sum of the numerators over the common denominator.
20. Some students write skillfully.
21. Some complex numbers are imaginary numbers.
22. None of us is perfect.
23. Only members are admitted to the club.

2.3 CONJUNCTION

The conjunction *and* is commonly used to combine sentences and phrases to form larger sentences. Similarly, the ampersand symbol "&" is used in logic to form the statement "$A \& B$" from the statements "A" and "B"; "$A \& B$" is read "A and B" or "the conjunction of A and B".

In translating into symbolic form, care must be taken that the symbols "A" and "B" are indeed translations of statements. For the statement

The number twelve is rational and positive,

a translation directly into symbols is not possible since the word "positive" is not a statement. If the statement is changed to the form

The number twelve is rational and
the number twelve is positive, then a direct translation is clearly "$A \& B$" where "A" and "B" are the translations:

A: The number twelve is rational.
B: The number twelve is positive.

If "P" and "Q" are statements, then we say that the conjunction "$P \& Q$" is a statement that is true in case both "P" and "Q" are true, and that in all other cases the conjunction is false.

From this definition it follows that the truth of "$P \& Q$" is dependent solely on the truth values of "P" and "Q", and not on any other relationships that might exist between "P" and "Q".

Normally, good English usage requires that the parts of a conjunction have some relation to each other. There appears to be no good reason for making a conjunction of the statements "Euclid is dead" and "Twelve is an integer" because they have no discernible relation beyond being both true. But to give precise criteria for distinguishing between acceptable and unacceptable conjunctions seems quite difficult. The problem is avoided in logic by permitting conjunction of any two statements. This latitude in the use of "&" permits some odd looking translations to occur, but does no harm. In practice, the use of "&" is quite parallel to the use of "and" in ordinary discourse. The situation with respect to the disjunction "or" is not so satisfactory.

2.4 DISJUNCTION

The disjunction "or" is used to connect two clauses or sentences to form a larger sentence. The meaning of this connection seems generally to be dependent on the meanings of the parts connected. For example,

He will succeed *or* die in the attempt.

A simple closed curve in the plane divides it into two regions such that any point not on the curve is either inside *or* outside the curve.

He'll have to fish *or* cut bait.

In these sentences the use of "or" is clearly meant to be exclusive; that is, the two assertions connected by "or" are to be considered mutually exclusive. This exclusive use of the word "or" as an indication of a clear alternative is the only one sanctioned by *Webster's New International Dictionary*,[1] but one can find many common examples of the nonexclusive use of "or".

Consider the following example taken from a sign on a private pond:

Residents or women may fish here without permits.

[1] *Webster's New International Dictionary of the English Language*, second edition, unabridged. Springfield, Mass.: G. & C. Merriam Company, 1952.

Does this sign mean what it seems to say? Would a person who is both a woman and a resident need a permit to fish? That would be the implication of the sign if its designer intended the "or" to be taken in the exclusive sense. This is clearly not the intent. In this sentence the word "or" is used in the inclusive sense. We might dismiss this as an example of poor English, since the meaning can be conveyed by the correct construction:

Residents and women may fish here without permits.

However, if the sign had read

Any person who is a resident or a woman
may fish here without a permit;

then, while the "or" is again used inclusively, it cannot be changed to "and" without radically altering the intended meaning of the sentence.

In the sentence

You will damage the motor if you run
it when too low on oil or water,

the intended meaning is conveyed by the inclusive "or". Again, to replace "or" by "and" would alter the intended meaning. Perhaps the meaning might best be conveyed in this case by using the awkward hybrid "and/or".

In the foregoing examples, the meaning of the "or" is clear enough, but its meaning is dependent on context. The ambiguity of "or" in the English language is not unique among Western languages. The French "ou", the German "oder", the Russian "ili" have much the same ambiguity—that is, they take their meaning partly from context. In Latin, "aut _____ aut _____" was used to indicate the exclusive "or", while "vel" was used to indicate the inclusive "or". In English, perhaps the closest we can come to the Latin "vel" is the awkward "and/or" sometimes used in commercial and legal documents. The unhappy construction "A or B, or both" is also used to convey the sense of the inclusive disjunction.

In a symbolic logic, a symbol for disjunction is needed that is independent of context. Because "or" is commonly used in the inclusive sense in mathematics, we choose this usage for the symbolic logic. The symbol "$\vee$", from the Latin "vel", is used to represent the inclusive "or".

If "P" and "Q" are statements, then "$P \vee Q$" is a statement that is true either when "P" is true or "Q" is true or both are true. "$P \vee Q$" is false only when both "P" and "Q" are false.

Suppose that "P" is a translation for "Euclid is dead" and "Q" is a translation for "The sum of 2 and 3 is 8". Then "$P \vee Q$" is a statement, and a true one if we agree that "P" is true. In ordinary conversation two such unrelated statements as "P" and "Q" are seldom combined in a disjunction. However, no harm is done in permitting such examples to appear in the logic, and thereby we avoid the problem of defining the closer relationship between the two parts of a disjunction that seems desirable in ordinary discourse.

It is not necessary to have a special symbol to translate the exclusive "or", since there are a number of ways to express this disjunction by means of symbols already introduced. Taking "A" and "B" as statements, form the expression

$$(A \vee B) \,\&\, (\sim(A \,\&\, B)). \tag{2.1}$$

First, "$A \vee B$" is a statement [definition of "v"]. "$A \& B$" is also a statement [definition of "&"]; therefore, "$\sim(A \& B)$" is a statement [definition of "$\sim$"]. Hence, finally, the expression (2.1) is a statement [definition of "&"]. Second, if we check the statement (2.1) with the definitions of the truth values of the compounds appearing in it, there are the following cases to consider:

a. If "A" and "B" are true, then "$A \vee B$" is true and "$\sim(A \& B)$" is false, so that the conjunction (2.1) is false.
b. If "A" and "B" are both false, then "$A \vee B$" is false and "$\sim(A \& B)$" is true, so that the conjunction (2.1) is again false.
c. If "A" and "B" are opposite in truth value, then "$A \vee B$" and "$\sim(A \& B)$" are both true, so that their conjunction (2.1) is true.

In summary, the expression (2.1) is a statement that is true if, and only if, one of the statements "A", "B" is true and the other is false. This is just the notion of an exclusive disjunction.

In the same way, it can be verified that the statement

$$[A \,\&\, (\sim B)] \vee [(\sim A) \,\&\, B]$$

also expresses an exclusive disjunction.

EXERCISES

1. Take the following symbolic translations:

P: x is an irrational number.
Q: x is a complex number.

Write a translation for each of the following statements:

a. x is not an irrational number.
b. x is not a complex number.

c. x is an irrational and complex number.
d. x is an irrational or a complex number.

2. Take the following symbolic translations:

R: ABC is a triangle.
S: ABC is equilateral.

Write a translation for each of the following statements:

a. ABC is not a triangle.
b. Not, ABC is equilateral.
c. ABC is a triangle or ABC is equilateral.
d. ABC is an equilateral triangle.
e. ABC is a triangle or ABC is equilateral, but not both.

For each of the following sentences, express as many symbolic translations as are necessary, as in Examples 1 and 2, and write a translation of the sentence. Translate "or" in the inclusive sense.

3. x is a real and complex number but is not irrational.
4. x is an irrational number or a rational number.
5. James is a married man.
6. y is not isosceles or y has two equal angles. (Here it is to be understood that "y" is the name of a triangle.)
7. y is not a triangle or y is not isosceles or y has two equal angles.
8. $a/\sin A = b/\sin B = c/\sin C$. (This is a common type of abbreviation in mathematics, and always represents a conjunction of statements of equality.)
9. It is not the case that, all roses are red and all violets are blue.
10. Some roses are not red or some violets are not blue.
11. It is not the case that x is a real number and x is an imaginary number.
12. x is not a real number or x is not an imaginary number.
13. y is a parallelogram or y is a rectangle.
14. It is not the case that y is a parallelogram or y is a rectangle.
15. y is not a parallelogram and y is not a rectangle.

2.5 THE CONDITIONAL

We cannot easily deal with the logical structure of mathematical arguments without some convenient symbolism for translating statements of the "if ____ then ____" type. Suppose that "x" and "y" represent certain angles and that "A" and "B" are translations as follows:

A: x and y have their sides parallel.
B: $x = y$.

Our symbolism must easily translate such statements as

(2.2) If x and y have their sides parallel, then $x = y$,

and

x and y having their sides parallel
implies that $x = y$.

We translate these conditional statements by "$A \rightarrow B$". We refer to "$\rightarrow$" as the *implication* or the *conditional* symbol, and read it "implies" or sometimes "only if". In the conditional "$A \rightarrow B$", "A" is called the *antecedent* and "B" is called the *consequent.*

We define formally: if "P" and "Q" are statements, then "$P \rightarrow Q$" is a statement; "$P \rightarrow Q$" is a true statement unless "P" is true and "Q" is false, in which case it is a false statement.

Consider this definition of the conditional in relation to the statement (2.2), which was translated "$A \rightarrow B$". Note that (2.2) is not intended as a general statement about all pairs of angles x and y, but is a statement about two particular angles whose names are "x" and "y". If the angles are as pictured in Fig. 1, then by the foregoing definition, "$A \rightarrow B$" is a true statement since both "A" and "B" are true statements. For the angles pictured in Fig. 2, the statement "$A \rightarrow B$" is also true by the

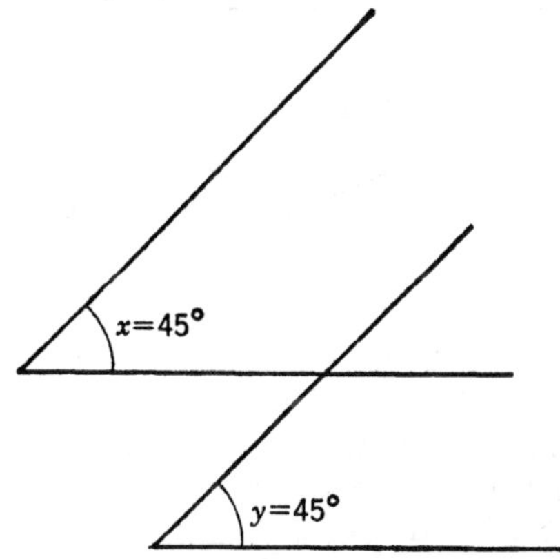

FIG. 1. Two angles of 45° with their sides parallel, left side to left side and right side to right side.

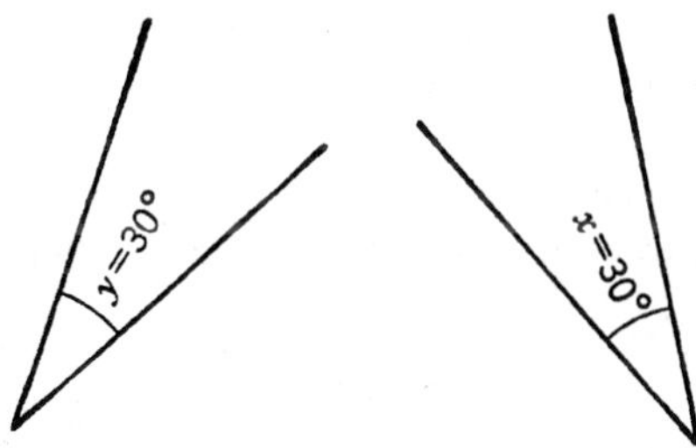

FIG. 2. Two angles x and y of 30°, but with their sides not respectively parallel.

definition, even though "A" is false and "B" is true. For angles as pictured in Fig. 3, "$A \rightarrow B$" is a false statement since "A" is true and "B" is false. Finally, for the angles pictured in Fig. 4, "$A \rightarrow B$" is a true statement by the definition, even though both "A" and "B" are false.

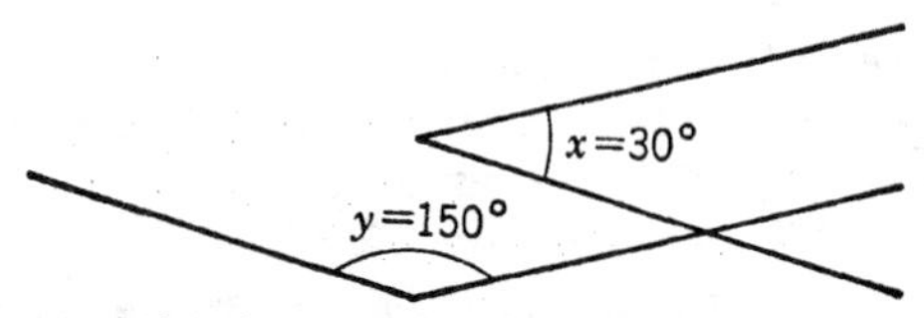

FIG. 3. Two angles $x = 30°$ and $y = 150°$ with their sides parallel left side to right side and right side to left side.

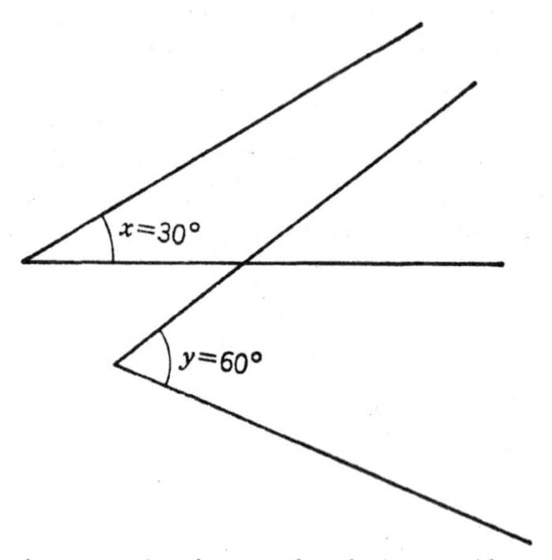

FIG. 4. Angles $x = 30°$ and $y = 60°$ such that no sides are parallel.

The definition of the conditional and the illustrations of its four cases are not in conflict with ordinary usage, but some of the true cases may appear strange because one does not ordinarily assert conditionals in which the antecedent is known to be false. However, as with the connectives previously considered, we want the truth value of the conditional to be defined for all the ways in which truth values can be assigned to its component statements. The unfamiliar cases may be regarded as merely supplementing ordinary usage of the conditional.

An example that makes the four cases of the conditional seem more comfortable comes from consideration of the following assertion. Your friend agrees,

If it is raining, then I shall drive you home.

We use the following translations.

P: It is raining.
Q: He drives you home.
$P \rightarrow Q$: If it is raining, then he drives you home.

The question is, under what conditions would you be angry with your friend?

If the situation is such that it is raining and he drives you home, then you are satisfied. In short, if "P" is true and "Q" is true, you are quite at ease with the notion that "$P \rightarrow Q$" is true. On the other hand, suppose it is not raining but your friend does drive you home. Would you have cause to be angry with him? No, because that was not the agreement in the first place. So you are still willing to accept "$P \rightarrow Q$" as true, even though "P" is false and "Q" is true. Now, suppose it is raining and he does not drive you home. Then you will be angry; you are quite willing to accept "$P \rightarrow Q$" as false in this case, "P" being true and "Q" being false. The fourth case, when it is not raining and your friend does not drive you home, leaves you with no cause for anger since that was the agreement all along. That is, you accept "$P \rightarrow Q$" as true (no anger) when "P" is false and "Q" is false.

It is particularly in the case of the conditional that one normally expects some relationship to exist between the antecedent and the consequent. In ordinary discourse the true conditional "(Euclid is dead) $\rightarrow (2 + 3 = 5)$" is ridiculous because of the lack of any sensible relationship between the death of Euclid and the sum of 2 and 3. Such conditionals are permitted in the logic for the same reason they are permitted in comparable combinations involving "&" and "v".

The latitude permitted in statements connected by "&", "v", and "$\rightarrow$" certainly makes these connectives weaker than the connectives "and", "or", and "implies" of ordinary discourse. Yet the logical connectives are strong enough for the purposes of mathematics, and have the advantage of precise definition, which the connectives of ordinary discourse do not.

The symbols so far introduced are adequate to deal with relations between mathematical statements as long as we do not attempt to investigate subject and predicate relations. Indeed, not all of "$\sim$", "&", "v", "$\rightarrow$" are needed. Frege[1] developed a statement calculus

[1] Gottlob Frege, *Begriffsschrift, eine der arithmetischen nachgebildete Formelsprache des reinen Denkens*, Halle: Nebert, 1879; or *Philosophical Writings of Gottlob Frege*, collected and translated by P. Geuch and M. Black, New York: Philosophical Library, Inc., 1952.

using just the symbols "$\rightarrow$" and "$\sim$", while Russell[1] did so using the symbols "$\sim$" and "v". Sheffer[2] gives an example of a statement calculus using a single symbol, not, however, equivalent to any symbol we have discussed. We shall continue to use all the symbols so far introduced, and perhaps even others, as a matter of convenience. To do so involves us in less complicated expressions and circumlocutions than would be the case if we limited ourselves to the symbols of Frege or Russell.

The dependence of the truth of compound statements upon the truth values of the statements occurring in them may be summarized in a truth table as follows:

(2.3)

A	B	$\sim A$	$A\&B$	$A \vee B$	$A \rightarrow B$
T	T	F	T	T	T
F	T	T	F	T	T
T	F	F	F	T	F
F	F	T	F	F	T

In the first two columns are listed the truth values assumed for the statements "A" and "B", while the succeeding columns list the corresponding truth values of the statements heading them. Of the four illustrations of the conditional (2.2), each corresponds to one of the rows of the truth table (2.3), where in this case the third, fourth, and fifth columns are ignored.

The connectives "&", "v", and "$\rightarrow$" are *binary* connectives; that is, they are used to combine statements two at a time. The connective "$\sim$" is always applied to a single statement, and is thus *singulary*. It is important to understand the dual nature of the definition of each of these symbols. In the first place, combining two statements by a binary connective produces a statement, and applying "$\sim$" to a statement produces a statement. In the second place, the truth of a binary compound is uniquely determined by the truth of its component statements, and the truth value of a negation is uniquely determined by the truth value of the statement negated. The truth table (2.3) summarizes this second aspect of the definitions.

2.6 STATEMENT FORMULAS

At this point, one cannot conclude that the expression "$P\&Q \vee R$" is a statement, even if it is known that "P", "Q", and "R" are statements.

[1] Bertrand Russell, *Introduction to Mathematical Philosophy*, pp. 144–150, New York: The Macmillan Company, 1919.

[2] H. M. Sheffer, "A Set of Five Independent Postulates for Boolean Algebras, with Application to Logical Constants," *Transactions of the American Mathematical Society*, vol. 14, pp. 481–488, 1913.

That "$(P\&Q)\vee R$" is a statement follows easily from the definitions. If "P" and "Q" are statements, then so is "$P\&Q$". The parentheses are used to indicate that "$(P\&Q)\vee R$" is a disjunction of the statements "$P\&Q$" and "R"; thus "$(P\&Q)\vee R$" is a statement. "$P\&Q\vee R$" can be regarded as unpunctuated. When it is punctuated by parentheses as indicated, it becomes a statement. Other punctuations are possible. "$P\&(Q\vee R)$" is a statement, but "$(P\&(Q\vee)R)$" and "$[P\&(Q]\vee R)$" are not.

An expression is considered well formed if it can be built up by a finite number of applications of the binary and singulary operations as indicated by the parentheses appearing in it.

Recall that the negation symbol is always applied to a single statement, so that while "$P\&[\sim(Q\vee R)]$" is properly formed, "$P\&[Q\sim(\vee R)]$" is not.

If it is desired to express the form of the statement "$(P\&Q)\vee R$" without any reference to the meanings of its component statements, one might try the expression "(_____&_____)v_____", where the blanks are looked upon as placeholders for statements. The expression is clearly not itself a statement, but may be regarded as the skeleton of a statement.

That there are drawbacks to this method of expressing the form of a statement by means of a skeleton is illustrated in the following example.

(2.4) John is married and Jane is his wife, or
John is married and Jane is his sister.

The statement (2.4) has for skeleton

(2.5) (_____&_____)v(_____&_____).

If the skeleton (2.5) truly expresses the form of (2.4), then the replacement of the blanks by any statements whatever should result in a statement having the same form as (2.4). What the skeleton does not indicate is the necessity of replacing the first and third blanks with the same statement in order to get a statement in the same form as (2.4). One might try to cope with this problem by using different colors, or numerical subscripts, for the blanks of the skeleton. On the whole, it seems simpler just to write

(2.6) $$(P\&Q)\vee(P\&R),$$

to indicate the form of (2.4) and to regard the letters "P", "Q", "R" not as translations of any of the component statements of (2.4), but merely as placeholders for any statements whatever.

The expression (2.6) is called a *statement formula.* It is not itself a statement, but it does indicate how a statement could be constructed from three given statements. In general, a statement formula is an expression composed of placeholder symbols, parentheses, and connective symbols that becomes a properly formed statement when the placeholders are replaced by properly formed statements.

If a placeholder "P" appears more than once in a statement formula, then in building a statement in the corresponding form, the same statement must replace "P" wherever it occurs. However, there is no requirement to replace different placeholders by different statements.

If, from three given statements, a statement is constructed in the form of

$$(P\&Q)\vee R, \tag{2.7}$$

then the truth of the resulting statement will usually depend on the truth values of the three given statements. The truth value can be computed with the help of the truth table (2.3). Suppose that the truth values of the statements replacing "P", "Q", "R" are, respectively, "F", "T", "F"; that is, *false, true, false.* From the second row of the truth table, it is found that "$P\&Q$" is replaced by a false statement. Hence, the constructed statement is a disjunction of a false statement and a false statement. Therefore it is false, as indicated by the fourth row of the truth table.

The truth value of any properly formed statement can be determined in the foregoing abbreviated manner. However, it is usually convenient to carry out such computations in tabular form. In the following truth table, the eight different ways truth values may be assigned in (2.7) are shown in the first three columns, and the last column shows the corresponding truth values assumed by (2.7).

P	Q	R	$P\&Q$	$(P\&Q)\vee R$
T	T	T	T	T
F	T	T	F	T
T	F	T	F	T
F	F	T	F	T
T	T	F	T	T
F	T	F	F	F
T	F	F	F	F
F	F	F	F	F

The sixth row of this truth table summarizes the analysis of (2.7) carried out in the preceding paragraph.

EXERCISES

Construct a complete truth table for each of the following statement formulas:

1. $R\&(\sim R)$
2. $\sim[R\&(\sim R)]$
3. $R\&(\sim S)$

4. $A \vee (\sim A)$
5. $P \& (Q \vee R)$; compare your table with the one on page 32.
6. $(P \to Q) \to R$
7. $(P \to Q) \to Q$; compare with Exercise 6 and with the table for $A \vee B$ in (2.3).
8. $(R \vee S) \& [\sim(R \& S)]$
9. $[R \& (\sim S)] \vee [(\sim R) \& S]$; compare with the table for Exercise 8.
10. $(P \& Q) \to P$

2.7 EQUIVALENCE

For the statement formula,

$$(P \to Q) \& (Q \to P), \tag{2.8}$$

the associated complete truth table is:

P	Q	$P \to Q$	$Q \to P$	$(P \to Q) \& (Q \to P)$
T	T	T	T	T
F	T	T	F	F
T	F	F	T	F
F	F	T	T	T

If "P" and "Q" are statements, then (2.8) is a statement. From the truth table we see that "$(P \to Q) \& (Q \to P)$" is true if "P" and "Q" have the same truth values and is false if "P" and "Q" have opposite truth values.

It is convenient to introduce "$P \leftrightarrow Q$" as an abbreviation for the statement formula (2.8). "$P \leftrightarrow Q$" is read "P equivalent Q". In table form we have:

P	Q	$P \leftrightarrow Q$
T	T	T
F	T	F
T	F	F
F	F	T

Note that, if for two statements "P" and "Q" it is known that "$P \leftrightarrow Q$" is true, it does not follow that the two statements are identical, or that they assert the same proposition. All that can be said is that the two statements have the same truth value.

With the symbolism already introduced, we can interpret some of the language used in mathematics. If "P" and "Q" are statements, we know that "$P \to Q$" is a statement. We have called "$P \to Q$" a conditional and have read it "P implies Q". There are other ways of reading the conditional that are often used in mathematical writing.

Some of these synonymous ways follow; remember each one would be translated as "$P \to Q$":

"Q" is a necessary condition for "P".
A necessary condition for "P" is "Q".
"P" is a sufficient condition for "Q".
A sufficient condition for "Q" is "P".

Clearly, if we have the assertion

"Q" is a necessary and sufficient condition for "P"

the translation is

$$P \leftrightarrow Q.$$

For, the first part,

"Q" is a necessary condition for "P",

assures us of the translation

$$P \to Q;$$

and the second part,

"Q" is a sufficient condition for "P",

yields the translation

$$Q \to P.$$

Together, we have

$$(P \to Q)\&(Q \to P),$$

which is another way to write

$$P \leftrightarrow Q.$$

EXERCISES

Write translations in terms of "P", "Q", "R" and the connectives for each of the following statements, taking as translations:

P: a is perpendicular to c.
Q: b is perpendicular to c.
R: a is parallel to b.

1. If a is perpendicular to c and b is perpendicular to c, then a is parallel to b.

2. If a is perpendicular to c and b is not perpendicular to c, then a is not parallel to b.

3. If a is perpendicular to c or b is perpendicular to c, then a is parallel to b or a is not parallel to b.

4. If a is not parallel to b, then a is not perpendicular to c or b is not perpendicular to c.

Using suitable translations "P", "Q", "R", . . . , write symbolic translations for each of the statements in Exercises 5 to 12.

5. If p and q are integers and $q \neq 0$, then p/q is a rational number.

6. If ABC is a triangle and ABC is isosceles, then ABC has two equal sides.

7. That ABC is a triangle and ABC has two equal sides is a necessary condition for it to be isosceles.

8. Whenever ABC is a triangle, then a sufficient condition for it to be isosceles is that it have two equal sides.

9. If a, b, c, and x are real numbers, $a \neq 0$, $ax^2 + bx + c = 0$, and $b^2 - 4ac = 0$, then the roots of $ax^2 + bx + c = 0$ are real and equal.

10. Whenever ABCD is a quadrilateral, then a necessary condition for it to be a square is that it be a rectangle.

11. Whenever ABCD is a quadrilateral, then a sufficient condition for it to be a square is that it be a rectangle whose sides are each 5 ft long.

12. Whenever ABCD is a quadrilateral, then a necessary and sufficient condition for it to be a square is that it have four right angles and four equal sides.

13. Express in symbols: The negation of a conditional is equivalent to the conjunction of its antecedent and the negation of its consequent.

14. Write out truth tables for "$(A \leftrightarrow B)$" and "$(A \vee B) \& (\sim(A \& B))$". Compare the truth tables. Refer back to (2.1) and read again the paragraph concerning it.

2.8 VALID STATEMENT FORMULAS

The class of statement formulas has an important subclass called the class of *valid statement formulas.* A valid statement formula has the property that whatever the truth values of the statements placed in it, the resulting statement is true; only formulas having this property are called *valid.*

The following formulas are examples of valid statement formulas:

$$A \vee (\sim A)$$
$$\sim(B \& (\sim B)).$$

To test the validity of a statement formula such as

$$[\sim(P \& Q)] \leftrightarrow [(\sim P) \vee (\sim Q)], \tag{2.9}$$

it is convenient to write out the complete truth table belonging to it.

P	Q	$P \& Q$	$\sim(P \& Q)$	$(\sim P) \vee (\sim Q)$	$[\sim(P \& Q)] \leftrightarrow [(\sim P) \vee (\sim Q)]$
T	T	T	F	F	T
F	T	F	T	T	T
T	F	F	T	T	T
F	F	F	T	T	T

From the last column of the truth table it follows that the statement formula (2.9) yields a true statement whatever the truth values of the statements placed in it at "P" and "Q". Hence, (2.9) is a valid statement formula.

As an illustration of the utility of formula (2.9) consider the problem of expressing a negation of the statement

(2.10) $\qquad l_1$ and l_2 are perpendicular to l_3,

where "l_1", "l_2", and "l_3" are names of three given lines. We can, of course, form the negative by prefixing "not" to (2.10), but the resulting statement is awkward. Instead let us change (2.10) into the conjunction

(2.11) $\qquad$ (l_1 is perpendicular to l_3)&(l_2 is perpendicular to l_3).

Now, the valid formula (2.9) shows us how to find a statement equivalent to the negation of any conjunction. Accordingly,

(2.12) $\quad$ (l_1 is not perpendicular to l_3)$\vee$(l_2 is not perpendicular to l_3)

is a statement equivalent to the negation of (2.11). The statement (2.12) is often more convenient to use than the negation of (2.10) or the negation of (2.11).

The statements in Exercises 9 to 12, Sec. 2.4, afford another illustration of the valid formula (2.9). The equivalence of these pairs of statements is just an instance of formula (2.9).

Because of their generality, the valid statement formulas occupy an important position in the statement logic. In effect, they describe *laws* of the logic in much the same way that trigonometric identities describe *laws* of trigonometry. The distinction between *equation* and *identical equation* in trigonometry is parallel to the distinction between *statement formula* and *valid statement formula* in the statement logic. For example, the equation $\sin\beta + \cos\beta = 1$ is a true statement of equality for some angles β (for example, $\beta = 0°, 90°, 360°, \ldots$) and is a false statement of equality for other angles β. However, the equation $\sin^2\beta + \cos^2\beta = 1$ is a true statement of equality for all angles β, and so describes a general property of the sine and cosine functions.

Certain of the valid statement formulas are formally analogous to algebraic identities expressing properties of real numbers. They are

(2.13)

a.	$(P\&Q) \leftrightarrow (Q\&P)$	Commutation, "&"
b.	$(P \vee Q) \leftrightarrow (Q \vee P)$	Commutation, "v"
c.	$[(P\&Q)\&R] \leftrightarrow [P\&(Q\&R)]$	Association, "&"
d.	$[(P \vee Q) \vee R] \leftrightarrow [P \vee (Q \vee R)]$	Association, "v"
e.	$[P\&(Q \vee R)] \leftrightarrow [(P\&Q) \vee (P\&R)]$	Distribution, "&" over "v"
f.	$[P \vee (Q\&R)] \leftrightarrow [(P \vee Q)\&(P \vee R)]$	Distribution, "v" over "&"

The first five statement formulas of (2.13) correspond to the following algebraic statements which are true for any real numbers "x", "y", "z":

a.	$x \cdot y = y \cdot x$	Commutation, multiplication
b.	$x + y = y + x$	Commutation, addition
c.	$(x \cdot y) \cdot z = x \cdot (y \cdot z)$	Association, multiplication
d.	$(x + y) + z = x + (y + z)$	Association, addition
e.	$x \cdot (y + z) = x \cdot y + x \cdot z$	Distribution, multiplication over addition

With the analogy as drawn above, there fails to be any analogue for (*f*), for while it is the case, as in (*e*), that $2 \cdot (3 + 4) = 2 \cdot 3 + 2 \cdot 4$, it is of course not the case, as in (*f*), that $2 + (3 \cdot 4) = (2 + 3) \cdot (2 + 4)$.

The valid statement formula (2.13*c*) can be made the basis for a useful abbreviation. The expression "$P\&Q\&R$" is not a well-formed formula (Sec. 2.6) but may be taken as an abbreviation both for "$(P\&Q)\&R$" and for "$P\&(Q\&R)$". Of course, with this agreement, "$P\&Q\&R$" is ambiguous, but in view of the validity of (2.13*c*), the ambiguity can do no harm. Clearly (2.13*d*) can be made the basis for a similar abbreviation.

The expression "$P \rightarrow Q \rightarrow R$" should not be used as an abbreviation because the binary connective "$\rightarrow$" is not associative. That is,

(2.14) $$[(P \rightarrow Q) \rightarrow R] \leftrightarrow [P \rightarrow (Q \rightarrow R)]$$

is not a valid statement formula. Let us interpret (2.14) by taking for "P", "Q", "R" the statements

(2.15)

P: ABC is a triangle.
Q: $\angle A > \angle B$.
R: BC > AC.

where A, B, and C are points in the plane, as in Fig. 5. Here AC is an arc of the circle whose diameter is AB; hence, the angle at A is considered to be equal to 90°. With this interpretation, "$P \rightarrow (Q \rightarrow R)$" becomes the statement

(2.16) If ABC is a triangle, then $\angle A > \angle B$ implies BC > AC.

Taken as a general statement, (2.16) is a theorem of plane geometry, but here we take it as a statement about the particular geometrical figure pictured in Fig. 5. Thus, the statements taken for "P", "Q", and "R" have the truth values "F", "T", "F", respectively. It follows that (2.16) is a true statement.

With the interpretation (2.15), "$(P \to Q) \to R$" becomes

(2.17) Whenever ABC is a triangle implies ∠A > ∠B, then BC > AC.

Now, ABC is not a triangle; so the statement "ABC is a triangle implies ∠A > ∠B" is true. It is not true that "BC > AC", and so (2.17) is a conditional with a true antecedent and a false consequent, and thus is

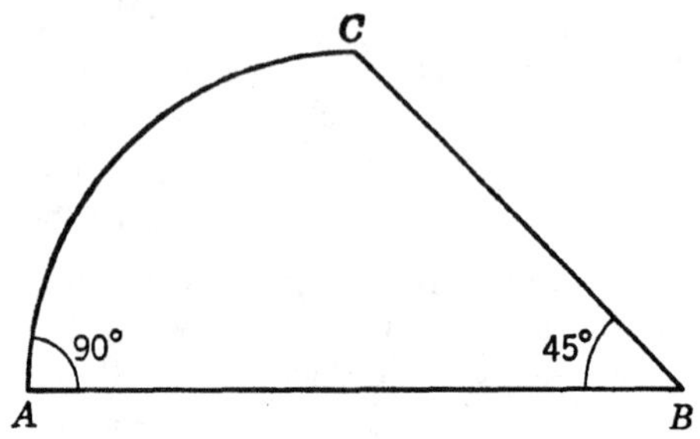

Fig. 5. AC is an arc of the circle whose diameter is AB.

false (see Sec. 2.5). Since under the interpretation (2.15) we have found (2.16) true and (2.17) false, it follows that (2.14) is not a valid statement formula. The complete truth table for (2.14) is:

P	Q	R	$P \to Q$	$Q \to R$	$(P \to Q) \to R$	$P \to (Q \to R)$	$[(P \to Q) \to R] \leftrightarrow [P \to (Q \to R)]$
T	T	T	T	T	T	T	T
F	T	T	T	T	T	T	T
T	F	T	F	T	T	T	T
F	F	T	T	T	T	T	T
T	T	F	T	F	F	F	T
F	T	F	T	F	F	T	F
T	F	F	F	T	T	T	T
F	F	F	T	T	F	T	F

Since the two properly formed formulas that can be formed out of "$P \to Q \to R$" by proper insertion of parentheses are not equivalent, this expression will not be employed as an abbreviation for anything.

2.9 NOTATIONAL CONVENTIONS

Parentheses may be omitted from formulas containing "&" alone, or "v" alone, but formulas containing both of these symbols generally require use of parentheses to avoid serious ambiguity. The same problem exists in arithmetic; $2 \cdot 3 + 4 \cdot 5$ might lead to the answers 50, 26, or 70, depending on whether the expression is viewed as meaning $[(2 \cdot 3) + 4] \cdot 5$, $(2 \cdot 3) + (4 \cdot 5)$, or $[2 \cdot (3 + 4)] \cdot 5$. Of course, the correct interpretation is the second one, leading to the answer 26. It is the correct interpretation because of the notational agreement made in arithmetic that in numerical statements involving both multiplication and addition, multiplication *binds* numbers more closely than addition.

A similar agreement can help to reduce the occurrence of parentheses in statement formulas. Let us agree that "&" binds statements more closely than "v". With this convention, we can take "$A \vee B \& C$" to be an abbreviation for "$A \vee (B \& C)$". We cannot expect to use "$A \vee B \& C$" also as an abbreviation for "$(A \vee B) \& C$" since, "$A \vee (B \& C) \leftrightarrow (A \vee B) \& C$" is not a valid statement formula.

In algebra, the product of x and y is usually written xy, without an explicit symbol to indicate the multiplication. This convention not only allows abbreviation of algebraic expressions but is also a practical aid in using the agreement on omission of parentheses. It is without doubt easier to interpret $xy + yz$ correctly than it is to interpret $x \cdot y + y \cdot z$ correctly.

To obtain similar benefits of easy interpretation for the statement logic, we use the abbreviation "AB" for "$A \& B$". Then the formula "$A \vee (B \& C)$", which was abbreviated to "$A \vee B \& C$", can now be further abbreviated to "$A \vee BC$". With these conventions, many statement formulas will be formally analogous to algebraic expressions, where "v", "&", "$\leftrightarrow$" correspond formally to "+", "$\cdot$", "=".

For the negation symbol "$\sim$" the convention is that it binds a statement more closely than either "&" or "v". Thus "$\sim P \vee Q$" is an abbreviation for "$(\sim P) \vee Q$". To express the denial of a disjunction, parentheses will always be required, as in "$\sim(P \vee Q)$". Similarly, "$\sim PQ$" is an abbreviation for "$(\sim P)Q$", and the denial of a conjunction will always require parentheses, as in "$\sim(PQ)$". We shall also write "$\sim\sim P$" for "$\sim(\sim P)$", and "$\sim\sim\sim P$" for "$\sim[\sim(\sim P)]$" from now on.

Finally, "$\sim$", "&", and "v" are taken to bind statements more closely than either "$\rightarrow$" or "$\leftrightarrow$". With this convention, "$P \vee Q \leftrightarrow Q \vee P$" is a legitimate abbreviation for "$(P \vee Q) \leftrightarrow (Q \vee P)$".

There will be no further conventions, so that if "$\rightarrow$" or "$\leftrightarrow$" are repeated, or both occur, in the same formula, parentheses will still be required to avoid ambiguity.

In summary, the notational conventions provide for eliminating the "&" symbol by using juxtaposition to indicate conjunction. The conventions provide for reducing the number of parentheses needed by ordering the connectives according to the closeness with which they bind statements, with "$\sim$" binding closest, then "&", then "v", and all of these binding statements more closely than "$\rightarrow$" or "$\leftrightarrow$". Thus to restore parentheses to the formula

$$A \vee B \sim C \rightarrow \sim D \vee F,$$

parentheses are placed first about the negation symbols and the statements to which they apply to obtain

$$A \vee B(\sim C) \rightarrow (\sim D) \vee F.$$

Next, parentheses are placed about the conjunctions

$$A \vee (B(\sim C)) \rightarrow (\sim D) \vee F,$$

and finally parentheses are placed about the disjunctions

$$(A \vee (B(\sim C))) \rightarrow ((\sim D) \vee F).$$

If desired, this last formula may be written with distinctive collection symbols

$$\{A \vee [B(\sim C)]\} \rightarrow \{(\sim D) \vee F\}.$$

With these conventions, formulas (2.13) can be written

$$(2.18) \quad \begin{aligned} &a.\ PQ \leftrightarrow QP \\ &b.\ P \vee Q \leftrightarrow Q \vee P \\ &c.\ (PQ)R \leftrightarrow P(QR) \\ &d.\ (P \vee Q) \vee R \leftrightarrow P \vee (Q \vee R) \\ &e.\ P(Q \vee R) \leftrightarrow PQ \vee PR \\ &f.\ P \vee QR \leftrightarrow (P \vee Q)(P \vee R) \end{aligned}$$

A person who feels at home with the symbolism of algebra will be quite comfortable with all of these formulas except (2.18*f*), which will seem strange because its algebraic counterpart is not an identity.

Almost all the formulas that follow have already been encountered in unabbreviated form and are here compared with their abbreviated forms:

Unabbreviated	*Abbreviated*
$[\sim(P \& Q)] \leftrightarrow [(\sim P) \vee (\sim Q)]$	$\sim(PQ) \leftrightarrow \sim P \vee \sim Q$
$(A \vee B) \& [\sim(A \& B)]$	$(A \vee B) \sim (AB)$
$[A \& (\sim B)] \vee [(\sim A) \& B]$	$A \sim B \vee \sim AB$
$[\sim(P \& Q)] \vee R$	$\sim(PQ) \vee R$
$[(\sim P) \& (\sim Q)] \vee R$	$\sim P \sim Q \vee R$
$(P \rightarrow Q) \& (Q \rightarrow P)$	$(P \rightarrow Q)(Q \rightarrow P)$
$[\sim(P \vee Q)] \leftrightarrow [(\sim P) \& (\sim Q)]$	$\sim(P \vee Q) \leftrightarrow \sim P \sim Q$

EXERCISES

1. Construct truth tables for the above formulas, and determine which of them are valid statement formulas.

2. Show by means of truth tables that the following are valid statement formulas:

a. $PQ \rightarrow P$

b. $(P \rightarrow A\vee B) \leftrightarrow [\sim B \rightarrow (P \rightarrow A)]$

c. $(P \rightarrow A\vee B) \leftrightarrow [(P \rightarrow \sim B) \rightarrow (P \rightarrow A)]$

d. $(A \rightarrow Q)(B \rightarrow Q) \leftrightarrow (A\vee B \rightarrow Q)$

e. $[P \rightarrow (Q \rightarrow R)] \leftrightarrow [PQ \rightarrow R]$

f. $(P \rightarrow A\vee B) \leftrightarrow (P \rightarrow A)\vee(P \rightarrow B)$

g. $(P \rightarrow Q)(P \rightarrow R) \leftrightarrow (P \rightarrow QR)$

h. $\sim\sim P \leftrightarrow P$

i. $\sim(PB) \leftrightarrow (P \rightarrow \sim B)$

3. As was mentioned in Sec. 2.5, not all of the connectives "$\sim$", "&", "v", "$\rightarrow$", "$\leftrightarrow$" are essential to the logic. For example, we could take "$\sim$" and "&" as basic symbols, and define the other connectives in terms of these. Recall that "$P\vee Q$" is true except when both "P" and "Q" are false. This suggests expressing "$P\vee Q$" in the form "$\sim(\sim P\sim Q)$". It is then easy to check by truth tables that "$P\vee Q \leftrightarrow \sim(\sim P\sim Q)$" is a valid statement formula.

a. Using only the symbols: "$\sim$", "&", "P", "Q", "(", ")", find formulas equivalent to "$P\vee Q$", "$P \rightarrow Q$", "$P \leftrightarrow Q$".

b. Using only the symbols: "$\sim$", "v", "P", "Q", "(", ")", find formulas equivalent to "$P\&Q$", "$P \rightarrow Q$", "$P \leftrightarrow Q$".

2.10 REPLACEMENT

It is often convenient to rename some or all of the placeholders in a statement formula. If in

$$A\vee(B \rightarrow C) \tag{2.19}$$

"A" is replaced by "P" and "B" by "Q", then (2.19) becomes

$$P\vee(Q \rightarrow C). \tag{2.20}$$

While (2.19) and (2.20) are not equivalent in the sense of Sec. 2.7, they are, nevertheless, alike in the sense that any statement that can be derived from (2.19) by replacing its placeholders by statements can also be derived from (2.20) by a suitable replacement of placeholders. Conversely, any statement derivable from (2.20) can also be derived from (2.19). To put it in another way, the totality of statements in the form of (2.19) is identical with the totality of statements in the form of (2.20).

Renaming the placeholders of a formula does not always yield a formula which is the same in the sense of the foregoing paragraph. If, for instance, "A" is replaced by "P" and "B" by "C" in (2.19) the result is

$$P \vee (C \rightarrow C). \tag{2.21}$$

The disjunction (2.21) is valid since "$C \rightarrow C$" is true for any statement placed in "C". Since (2.19) is clearly not a valid formula, there is at least one statement derivable from (2.19) that cannot be derived from (2.21). However, any statement derivable from (2.21) can be derived from (2.19). That is to say, the totality of statements in the form of (2.21) is a subset of the totality of statements in the form of (2.19).

We formulate the notion of replacement in somewhat more general terms in the following, where we use script letters as names of formulas.

Replacement Rule: A formula $\mathcal{G}$ is said to be derived from a formula $\mathcal{F}$ *by replacement of* "A" *by* $\mathcal{B}$ if and only if *every* occurrence of the placeholder "A" in $\mathcal{F}$ is replaced by the same formula $\mathcal{B}$.

Note that the symbol replaced must be a placeholder, but that it may be replaced by a statement formula. For instance, from "$A \vee \sim A$" we can obtain "$(P \rightarrow Q) \vee \sim (P \rightarrow Q)$" by replacing the placeholder "A" by the formula "$P \rightarrow Q$".

With this more general kind of replacement, it is still the case that if $\mathcal{G}$ is obtained from $\mathcal{F}$ by replacement, then every statement derivable from $\mathcal{G}$ is also derivable from $\mathcal{F}$. It follows that if $\mathcal{F}$ is a valid formula, then so is $\mathcal{G}$ a valid formula. It is this property of the replacement process that makes it a useful tool. For instance, in the valid formula, Exercise 2*e*, Sec. 2.9,

$$[P \rightarrow (Q \rightarrow R)] \leftrightarrow [PQ \rightarrow R],$$

"R" may be replaced by "P" to obtain the valid formula

$$[P \rightarrow (Q \rightarrow P)] \leftrightarrow [PQ \rightarrow P].$$

It should be observed that to obtain (2.20) from (2.19) requires two applications of the replacement rule.

2.11 SUBSTITUTION

In Sec. 2.8 the method of truth tables was used to establish the validity of formula (2.9):

$$\sim (PQ) \leftrightarrow \sim P \vee \sim Q.$$

By replacing "P" by "$\sim P$" and "Q" by "$\sim Q$", the valid formula

$$(2.22) \qquad \sim(\sim P \sim Q) \leftrightarrow \sim\sim P \vee \sim\sim Q$$

is obtained. By Exercise 2*h*, Sec. 2.9,

$$(2.23) \qquad \sim\sim P \leftrightarrow P$$

is valid. Also, by replacement in (2.23),

$$(2.24) \qquad \sim\sim Q \leftrightarrow Q$$

is valid. On considering these last three formulas, it seems reasonable to expect that

$$(2.25) \qquad \sim(\sim P \sim Q) \leftrightarrow P \vee Q$$

is a valid formula. Truth table analysis will show that it is indeed valid.

The replacement rule will not yield (2.25) from (2.22) since it would be necessary to replace formula "$\sim\sim P$" by "P" and formula "$\sim\sim Q$" by "Q", which would violate the replacement rule requirement that placeholders be replaced rather than formulas. Yet in view of the equivalences (2.23) and (2.24) such substitutions seem reasonable, and it appears likely that a suitably formulated principle of substitution might well enable one to establish validity in some cases without resorting to truth tables. Again we use script letters as names of formulas and write

Substitution Principle: If $\mathcal{U}$ is a formula containing a component formula $\mathcal{A}$, and if $\mathcal{U}^*$ is a formula obtained from $\mathcal{U}$ by substituting a formula $\mathcal{B}$ for one or more occurrences of $\mathcal{A}$ in $\mathcal{U}$, and if $\mathcal{A} \leftrightarrow \mathcal{B}$ is valid, then $\mathcal{U} \leftrightarrow \mathcal{U}^*$ is valid.

An example of the use of the substitution principle follows. Suppose

$\mathcal{U}$ denotes $\sim(PQ) \rightarrow P$ [This formula is, of course, not valid.]
$\mathcal{A}$ denotes $\sim(PQ)$
$\mathcal{B}$ denotes $\sim P \vee \sim Q$

On substituting $\mathcal{B}$ for $\mathcal{A}$ in $\mathcal{U}$, we obtain for $\mathcal{U}^*$

$$\sim P \vee \sim Q \rightarrow P.$$

Now since "$\sim(PQ) \leftrightarrow \sim P \vee \sim Q$" is valid (formula 2.9), we have $\mathcal{A} \leftrightarrow \mathcal{B}$ is valid and it follows by the substitution principle that $\mathcal{U} \leftrightarrow \mathcal{U}^*$ or

$$[\sim(PQ) \rightarrow P] \leftrightarrow [\sim P \vee \sim Q \rightarrow P]$$

is a valid formula. Truth-table analysis will verify the validity, but is not needed.

A valuable result is obtained by applying the substitution principle to a valid formula. Suppose $\mathfrak{A}$ is a valid formula, and $\mathfrak{A}^*$ is obtained from $\mathfrak{A}$ by use of the substitution principle. Then

$$\mathfrak{A} \leftrightarrow \mathfrak{A}^*$$

is valid. Since every assignment of truth values to the placeholders of $\mathfrak{A}$ results in the value "T" for $\mathfrak{A}$, then the same must be true for $\mathfrak{A}^*$, and it follows that $\mathfrak{A}^*$ is a valid statement formula. In short, substitution in a valid formula yields a valid formula. It is exactly this result that is needed to obtain (2.25) from (2.22) by substitution from (2.23) and (2.24).

In considering the quite distinct notions of *replacement* and *substitution*, it is well to emphasize the following points:

"Replacement" is a means of renaming placeholders and

a. Only placeholders may be replaced.
b. If a placeholder is to be replaced, it must be replaced in every one of its occurrences.

"Substitution" is a means of obtaining essentially new formulas and

a. For any formula a validly equivalent formula may be substituted.
b. It is not necessary to substitute in all occurrences of the formula in question.

EXERCISES

In the following exercises, use the replacement rule and the substitution principle to show validity:

1. Show that "$\sim\sim[\sim(P \rightarrow Q)] \leftrightarrow \sim(P \rightarrow Q)$" is valid using "$\sim\sim P \leftrightarrow P$" (Exercise 2*h*, Sec. 2.9).

2. Show that "$\sim\sim(R\sim R) \leftrightarrow R\sim R$" is valid.

3. Show that "$P \rightarrow (Q \rightarrow P)$" is valid, using Exercises 2*a* and 2*e*, Sec. 2.9.

4. Start with the valid formula of Exercise 2*d*, Sec. 2.9, and obtain the valid formula "$(\sim A \rightarrow Q)(\sim B \rightarrow Q) \leftrightarrow [\sim(AB) \rightarrow Q]$". You will need (2.9).

5. Obtain the valid formula "$PQ \leftrightarrow \sim(\sim P \vee \sim Q)$". Start with "$\sim\sim P \leftrightarrow P$", use replacement, and substitution from (2.9).

6. Obtain the valid formula "$\sim(P \rightarrow Q) \leftrightarrow P\sim Q$". Start with Exercise 2*h*, Sec. 2.9, and use Exercise 2*i*.

2.12 SOME USEFUL VALID FORMULAS

Formula (2.25):

$$\sim(\sim P \sim Q) \leftrightarrow P \vee Q,$$

which can be shown to be valid by the methods of Sec. 2.10 and 2.11, can be regarded as showing how to express "v" in terms of "$\sim$" and "&". Formula (2.26), which follows, can be regarded as showing how to express "$\rightarrow$" in terms of "$\sim$" and "v".

$$(P \rightarrow Q) \leftrightarrow \sim P \vee Q. \tag{2.26}$$

A proof for (2.26) is given in the following truth table:

P	Q	$P \rightarrow Q$	$\sim P \vee Q$	$(P \rightarrow Q) \leftrightarrow \sim P \vee Q$
T	T	T	T	T
F	T	T	T	T
T	F	F	F	T
F	F	T	T	T

If in (2.26) "P" is replaced by "$\sim P$" and then "P" substituted for "$\sim\sim P$", a valid formula is obtained which may be regarded as an alternative form of (2.26). It is

$$(\sim P \rightarrow Q) \leftrightarrow P \vee Q. \tag{2.27}$$

Suppose "P" and "Q" are translations of statements as follows:

P: You are completely satisfied.
Q: You get your money back.

Then "$P \vee Q$" is a translation of

$P \vee Q$: You are completely satisfied or you get your money back.

By (2.27), an equivalent statement is

$\sim P \rightarrow Q$: If you are not completely satisfied, then you get your money back.

Perhaps a manufacturer making such a guarantee has in mind the exclusive "or", so that "$P \vee Q$" is not a translation of his guarantee. However, following the doctrine that "the customer is always right", he will not quarrel with customer claims and will have to pay off as if the guarantee is given with the inclusive "or".

For a trigonometric interpretation of (2.26), let "P" and "Q" be the translations

P: $\theta + \phi = 90°$.
Q: $\sin \theta = \cos \phi$.

Then the interpretation of (2.26) is

$$[(\theta + \phi = 90°) \rightarrow (\sin \theta = \cos \phi)] \leftrightarrow [(\theta + \phi \neq 90°)\text{v}(\sin \theta = \cos \phi)].$$

Note that it is quite possible for both "$\theta + \phi \neq 90°$" and "$\sin \theta = \cos \phi$" to be true statements. (This will be the case, for instance, if $\theta = \phi = 225°$.) For this interpretation, "$P \rightarrow Q$" is known to be true for all angles θ and ϕ; thus by (2.26) "$\sim P\text{v}Q$" is also true for all angles θ and ϕ.

A companion to the valid formula (2.9)

$$\sim(PQ) \leftrightarrow \sim P\text{v}\sim Q, \tag{2.28}$$

is the valid formula

$$\sim(P\text{v}Q) \leftrightarrow \sim P\sim Q. \tag{2.29}$$

Formulas (2.28) and (2.29) are known as De Morgan's laws after the English logician A. De Morgan (1806–1878), though these forms were known and used long before his time.

The validity of (2.29) can be established by truth tables, but it is instructive to do so by means of replacement and substitution. Replace "P" by "$\sim P$", and "Q" by "$\sim Q$" in (2.28) to obtain

$$\sim(\sim P\sim Q) \leftrightarrow \sim\sim P\text{v}\sim\sim Q.$$

Substitution for "$\sim\sim P$" and "$\sim\sim Q$" yields

$$\sim(\sim P\sim Q) \leftrightarrow P\text{v}Q. \tag{2.30}$$

Now, following the pattern of the Substitution Principle, take for $\mathfrak{U}$

$$\sim(P\text{v}Q),$$

and substitute from (2.30) to obtain for $\mathfrak{U}^*$

$$\sim\sim(\sim P\sim Q).$$

By the Substitution Principle, "$\mathfrak{U} \leftrightarrow \mathfrak{U}^*$"; so

$$\sim(P\text{v}Q) \leftrightarrow \sim\sim(\sim P\sim Q),$$

and a final substitution to remove the double negation yields

$$\sim(P\text{v}Q) \leftrightarrow \sim P\sim Q,$$

which is the required formula (2.29).

A theorem of mathematics is commonly stated in the form of a conditional. If "$P \rightarrow Q$" is the translation of a theorem, it is sometimes easiest to prove it indirectly by showing that "$\sim(P \rightarrow Q)$" is a false statement. If "$\sim(P \rightarrow Q)$" is false, then by definition of negation, "$P \rightarrow Q$" is true. Unfortunately, "$\sim(P \rightarrow Q)$" is grammatically awkward in translation. It is useful to have an equivalent form for

"$\sim(P \rightarrow Q)$" which is more comfortable in translation. Such a formula can be derived by the methods of Secs. 2.10 and 2.11.

Start with

$$\sim(P \rightarrow Q)$$

and substitute from (2.26) to get

$$\sim(P \rightarrow Q) \leftrightarrow \sim(\sim P \vee Q). \tag{2.31}$$

If "Q" is replaced by "$\sim Q$" in the De Morgan formula (2.28), and if then "Q" is substituted for "$\sim\sim Q$", the result is

$$\sim(P\sim Q) \leftrightarrow \sim P \vee Q.$$

This is then substituted in (2.31) to get

$$\sim(P \rightarrow Q) \leftrightarrow \sim\sim(P\sim Q).$$

A final substitution to remove the double negation yields the desired valid formula

$$\sim(P \rightarrow Q) \leftrightarrow P\sim Q. \tag{2.32}$$

Suppose we wish to prove the statement

(2.33) If a $\perp$ c and b $\perp$ c, then a $\|$ b,

where "a", "b", "c" are unknown, but fixed, lines in a plane. With translations,

P: a $\perp$ c and b $\perp$ c
Q: a $\|$ b,

a translation of (2.33) is "$P \rightarrow Q$". To prove "$P \rightarrow Q$" indirectly, we should have to start with

(2.34) $\sim(P \rightarrow Q)$: Not; if a $\perp$ c and b $\perp$ c, then a $\|$ b,

which is in an awkward form. By (2.32) a form equivalent to (2.34) is

(2.35) $P\sim Q$: a $\perp$ c and b $\perp$ c, and a is not parallel to b.

The form of (2.35) is quite comfortable to work with. If it is proved false, then so is (2.34) false, and (2.33) is proved true.

EXERCISES

First, write the negation of each of the following statements in a form suitable for direct translation into symbols. That is, for "The eye is black and blue", write, "Not; the eye is black and the eye is blue". Second, if the resulting negation seems grammatically awkward, use valid formulas, replacement, or substitution to find an equivalent statement that is more graceful. That is, the above statement would be changed to, "The eye is not black or the eye is not blue". (Where "or" appears, take it in the inclusive sense.)

1. I shall not be home Tuesday or Wednesday.
2. The bus stops at Bay Street or Water Street.
3. My office hours are on Tuesday, Wednesday, and Saturday.
4. m or n is a zero of $f(x)$.
5. No refund or exchange is permitted.
6. It is not true that John will go to college or to work upon graduation from high school.
7. $\triangle$ABC is isosceles or equilateral.
8. The pay telephone takes nickels, dimes, or quarters.
9. If the weather is mild this fall, people will not buy heavy coats.
10. If $a \neq b$ and $c \neq d$, then $a + c \neq b + d$.
11. If $a > b$ and $b > c$, then $a > c$.
12. If two triangles are congruent, their corresponding sides are equal.
13. $a > b$.
14. a $\perp$ c and b $\perp$ c, and a is not parallel to b.
15. a $\|$ b, and c is a transversal forming alternate interior angles α and β with a and b, and $\alpha \neq \beta$.
16. If n is an even number, then n is exactly divisible by 2.
17. If $\sqrt{2}$ is irrational, it cannot be expressed in the form p/q, $q \neq 0$, where p and q are integers.
18. $\sqrt{2}$ is irrational.
19. If n is rational, it can be expressed as the quotient of two integers, p/q, $q \neq 0$.
20. If $P(x)$ is a polynomial with real coefficients, then whenever $a + bi$ is a root of $P(x)$ so is $a - bi$ a root of $P(x)$. (Translate in the form: "$A \rightarrow (B \rightarrow C)$".)

2.13 A COLLECTION OF VALID STATEMENT FORMULAS

Below is listed a group of valid statement formulas that are important in their own right or are referred to one or more times in the sequel.

	Formula	*Descriptive name*	*Proved in section*
(2.36)	$A \vee \sim A$	Law of the excluded middle	
(2.37)	$\sim(A \sim A)$	Law of contradiction	
(2.38)	$\sim(PQ) \leftrightarrow \sim P \vee \sim Q$	De Morgan's law	2.8
(2.39)	$\sim(P \vee Q) \leftrightarrow \sim P \sim Q$	De Morgan's law	2.12
(2.40)	$\sim\sim P \leftrightarrow P$	Double negative	

(2.41)	$\sim(P \to Q) \leftrightarrow P{\sim}Q$	Negation of a conditional	2.12
(2.42)	$(P \to Q) \leftrightarrow (\sim Q \to \sim P)$	Contrapositive	2.14
(2.43)	$(P \to Q) \leftrightarrow \sim P \mathrm{v} Q$	Equivalent form of the conditional	2.12
(2.44)	$PQ \leftrightarrow QP$	Commutative law	
(2.45)	$P \mathrm{v} Q \leftrightarrow Q \mathrm{v} P$	Commutative law	
(2.46)	$(PQ)R \leftrightarrow P(QR)$	Associative law	
(2.47)	$(P \mathrm{v} Q) \mathrm{v} R \leftrightarrow P \mathrm{v} (Q \mathrm{v} R)$	Associative law	
(2.48)	$P(Q \mathrm{v} R) \leftrightarrow PQ \mathrm{v} PR$	Distributive law	
(2.49)	$P \mathrm{v} QR \leftrightarrow (P \mathrm{v} Q)(P \mathrm{v} R)$	Distributive law	
(2.50)	$PQ \to P$		2.9
(2.51)	$PQ \to Q$		2.9
(2.52)	$(P \to A \mathrm{v} B) \leftrightarrow [\sim B \to (P \to A)]$		2.9
(2.53)	$(P \to A \mathrm{v} B) \leftrightarrow [(P \to \sim B) \to (P \to A)]$		2.9
(2.54)	$(A \to Q)(B \to Q) \leftrightarrow (A \mathrm{v} B \to Q)$		2.9
(2.55)	$[P \to (Q \to R)] \leftrightarrow (PQ \to R)$		2.9
(2.56)	$(P \to A \mathrm{v} B) \leftrightarrow (P \to A) \mathrm{v} (P \to B)$		2.9
(2.57)	$(P \to Q)(P \to R) \leftrightarrow (P \to QR)$		2.9
(2.58)	$\sim(PB) \leftrightarrow (P \to \sim B)$		2.9
(2.59)	$(P \to Q)(Q \to R) \to (P \to R)$		2.14
(2.60)	$PQ \to (P \to Q)$		2.15

2.14 RULES OF INFERENCE

From the nature of their definition, valid statement formulas are laws of the statement logic. As such, they are useful in formal mathematical proofs. Fully to exploit these logical laws, some rules are needed for obtaining true statements from given statements, assumed or proved to be true. The most fundamental of these rules occurs tacitly in many geometric proofs of elementary mathematics. It is almost always required when a previously proved theorem is to be used as a step of a proof.

Consider the usual method of employing the congruence theorem commonly denoted by "S.S.S.". An instance of this theorem may be written

$$\text{(ABC is a triangle)(A}'\text{B}'\text{C}'\text{ is a triangle)(AB} = \text{A}'\text{B}')\text{(BC} = \text{B}'\text{C}')\text{(CA} = \text{C}'\text{A}') \to \text{(ABC} \cong \text{A}'\text{B}'\text{C}'). \tag{2.61}$$

The statement (2.61) is accepted as true because it is an instance of a general statement that has been proved true for all pairs of triangles. In the course of a proof requiring this theorem, it is customary first to prove in some way the statement

$$\text{(ABC is a triangle)(A}'\text{B}'\text{C}'\text{ is a triangle)(AB} = \text{A}'\text{B}')\text{(BC} = \text{B}'\text{C}')\text{(CA} = \text{C}'\text{A}'), \tag{2.62}$$

and then to infer the statement

(2.63) $$\text{ABC} \cong \text{A'B'C'}.$$

In the traditional two-column form of demonstration, the statements (2.62) and (2.63) appear in the left-hand column headed *Statements* (perhaps with " $\therefore$ " preceding (2.63)), and statement (2.61), or commonly just "S.S.S." appears in the right-hand column headed *Reasons*. Whatever the particular format, the result is the inference that (2.63) is considered proved because (2.61) and (2.62) are proved statements.

In the translation

P: Statement (2.62)
Q: Statement (2.63),

the inference takes the form

(2.64) If "P" and "$P \rightarrow Q$" are proved statements, then "Q" is inferred to be a proved statement.

Considering (2.64) as a general assertion about statements, we shall assume it as a rule of inference. It is often called the "Rule of Modus Ponens" or the "Law of Detachment", and we shall refer to it simply as "modus ponens" or sometimes just "mod pon". In (2.64), "$P \rightarrow Q$" is called the *major premise*, "P" the *minor premise*, and "Q" the *conclusion*.

As a convenient abbreviation for inference rules, we shall adopt a notation that for (2.64) takes the form

$$\frac{P \rightarrow Q, P}{Q} \quad \text{or} \quad \frac{P, P \rightarrow Q}{Q}, \qquad \text{mod pon}$$

We must distinguish carefully between the notion of a conditional and the notion of an inference by modus ponens. By itself, "$P \rightarrow Q$" is a formula whose truth values depend in a certain way upon the truth values of "P" and "Q". Even if it is known that "$P \rightarrow Q$" is true, nothing can be said about the truth of "Q". One cannot, so to speak, *detach* "Q" as true from a true "$P \rightarrow Q$". However, from a true "P" and a true "$P \rightarrow Q$", one can *detach* a true "Q" by modus ponens.

Note that from a consideration of the definition of "$\leftrightarrow$", it follows that if "P" and "$P \leftrightarrow Q$" are proved, then so is "Q" proved. We can regard this as a special case of modus ponens.

In the application of modus ponens to (2.61) it is necessary to know that (2.62) is true or proved. Customary procedure is to establish the truth of each of the five statements of the continued conjunction (2.62) separately, and then to *infer* from this the truth of (2.62) itself. This procedure depends upon a generalization of the following rule of inference:

(2.65) If "P" is proved and "Q" is proved, then "PQ" is inferred to be proved.

If (2.62) is translated without using the convention about omission of parentheses, it has the form "$((((PQ)R)S)T)$". Four applications of (2.65) are necessary to infer the truth of (2.62) from the truths of each of "P", "Q", "R", "S", "T".

The rule (2.65) will be called the *Rule of Conjunctive Inference*, and may be abbreviated

$$\frac{P,Q}{PQ}, \qquad \text{conj inf}$$

A rule that is related to conjunctive inference is often needed in proofs. It is called *conjunctive simplification:*

(2.66) If "PQ" is proved, then either "P" or "Q" is inferred to be proved.

Through use of (2.66) it is possible to "detach" from a conjunction any one of the statements in the conjunction. The rule (2.66) may be abbreviated

$$\frac{PQ}{P}, \text{ or } \frac{PQ}{Q}, \qquad \text{conj simp}$$

To derive the rule of conjunctive simplification, we use modus ponens with (2.50), or (2.51), as major premise and "PQ" as minor premise; then "P", or "Q", is inferred to be true.

The three rules of inference in conjunction with valid statement formulas provide additional means for inferring new statements from given statements. The valid statement formula

$$(P \rightarrow Q)(Q \rightarrow R) \rightarrow (P \rightarrow R)$$

is listed as (2.59). It embodies an important logical law, and its validity is established by the following truth table:

P	Q	R	$P \rightarrow Q$	$Q \rightarrow R$	$(P \rightarrow Q)(Q \rightarrow R)$	$P \rightarrow R$	$(P \rightarrow Q)(Q \rightarrow R) \rightarrow (P \rightarrow R)$
T	T	T	T	T	T	T	T
F	T	T	T	T	T	T	T
T	F	T	F	T	F	T	T
F	F	T	T	T	T	T	T
T	T	F	T	F	F	F	T
F	T	F	T	F	F	T	T
T	F	F	F	T	F	F	T
F	F	F	T	T	T	T	T

Suppose it is desired to prove a statement in the form "$P \rightarrow R$". If it is first possible to prove the two statements "$P \rightarrow Q$" and "$Q \rightarrow R$", then by the rule of conjunctive inference (2.65), "$(P \rightarrow Q)(Q \rightarrow R)$" may be inferred. After this, using "$(P \rightarrow Q)(Q \rightarrow R)$" as minor premise, and (2.59) as major premise, "$P \rightarrow R$" may be inferred by modus ponens. In summary, "$P \rightarrow R$" has been inferred from "$P \rightarrow Q$" and "$Q \rightarrow R$".

We can express this method of proving "$P \to R$" as a new inference rule derived from the two previous inference rules. It is called the *hypothetical syllogism:*

(2.67) $$\frac{P \to Q,\ Q \to R}{P \to R}, \qquad \text{hyp syll}$$

The derived inference rule (2.67) occurs frequently in mathematical proofs, and it is easily extended to a chain of several conditionals:

$$\frac{P \to Q,\ Q \to R,\ \ldots,\ S \to T}{P \to T}.$$

In the same way, a derived rule of inference stems from the valid formula (2.42):

(2.68) $$(P \to Q) \leftrightarrow (\sim Q \to \sim P).$$

The validity of (2.68) can be established by starting with the valid formula:

1. $(P \to Q) \leftrightarrow \sim P \vee Q$ [Formula (2.26)]
2. $(P \to Q) \leftrightarrow \sim P \vee \sim\sim Q$ [Substitution from "$Q \leftrightarrow \sim\sim Q$"]
3. $(A \to Q) \leftrightarrow \sim A \vee Q$ [Replace "P" by "A" in Step 1]
4. $(A \to \sim P) \leftrightarrow \sim A \vee \sim P$ [Replace "Q" by "$\sim P$" in Step 3]
5. $(\sim Q \to \sim P) \leftrightarrow \sim\sim Q \vee \sim P$ [Replace "A" by "$\sim Q$" in Step 4]
6. $(\sim Q \to \sim P) \leftrightarrow \sim P \vee \sim\sim Q$ [Substitution in Step 5 from (2.45) with replacement]
7. $(P \to Q) \leftrightarrow (\sim Q \to \sim P)$ [Substitution, Step 6 in Step 2]

Now then, if "$P \to Q$" is the translation of a proved statement, with it as minor premise, and (2.68) as major premise, "$\sim Q \to \sim P$" can be inferred to be true by modus ponens. The corresponding derived rule of inference is written

(2.69) $$\frac{P \to Q}{\sim Q \to \sim P}.$$

The formula "$\sim Q \to \sim P$" is called the *contrapositive* of "$P \to Q$", but the name *contrapositive inference* will be reserved for an extension of (2.69). Suppose both "$P \to Q$" and "$\sim Q$" are proved. By (2.69), "$\sim Q \to \sim P$" can be inferred; then, with "$\sim Q \to \sim P$" as major premise and "$\sim Q$" as minor premise, "$\sim P$" can be inferred by modus ponens. It is this scheme of inference that we shall call the *Rule of Contrapositive Inference.* In the notation adopted, it is written

(2.70) $$\frac{P \to Q,\ \sim Q}{\sim P}, \qquad \text{contrap inf}$$

Contrapositive inference occurs frequently in everyday discourse as well as in mathematics. An instance of (2.70) is the argument

> If Mr. Doe is a communist, he associates with communists.
> But Mr. Doe doesn't associate with communists.
> Therefore, Mr. Doe isn't a communist.

The argument is valid whatever the truth or falsity of the statements in it may be.

Another instance is the cliché

If John's a Democrat, I'll eat my hat.

By his tone of voice, the speaker makes it clear that he is not going to eat his hat, and the inference is that "John is not a Democrat".

At a certain stage in the standard proof of the irrationality of the square root of 2, it is known that for a certain integer x, x^2 is even. To proceed with the proof, it is necessary to show that x is even. It is easy to show that

$$(x \text{ odd}) \rightarrow (x^2 \text{ odd}),$$

for if x is odd, it can be expressed in the form $2n + 1$ for some integer n. Then

$$(2n + 1)^2 = 4n^2 + 4n + 1 = 2(2n^2 + 2n) + 1$$

which is clearly odd. With this proved, the desired result follows by contrapositive inference:

$$\frac{(x \text{ odd}) \rightarrow (x^2 \text{ odd}),\ \sim(x^2 \text{ odd})}{\sim(x \text{ odd})}$$

So it is proved that x is not odd, that is, x is even.

In Sec. 2.7, several alternative ways to read "$P \rightarrow Q$" were discussed. There are two additional ways in which the conditional occurs in mathematical writing. One of these is

P only if Q.

The sense of the above statement seems to be: if we do not have Q, then we cannot have P. But "$\sim Q \rightarrow \sim P$" is just a translation of this statement. Since "$(\sim Q \rightarrow \sim P) \leftrightarrow (P \rightarrow Q)$", we have by substitution the desired interpretation.

Another way in which the conditional "$P \rightarrow Q$" occurs is in the form

Q if P.

There is no question that the interpretation of this form is "$P \rightarrow Q$".

Now, let us put the two forms together:

P if and only if Q.

In this form we have, in translation,

$$(P \to Q)(Q \to P),$$

the familiar form of "$P \leftrightarrow Q$". Notice that the form is exactly like the one we obtained, Sec. 2.7, in the discussion of necessary and sufficient conditions. Current writing in mathematics tends to use "if and only if" much more than "necessary and sufficient".

EXERCISES

For each of Exercises 1 to 10, cite all the inference schemes used, the tacit theorems if any, and show by rewriting the statements explicitly how the inference schemes are used.

1. $x = y$
$y = z$
$\therefore x = y$ and $y = z$

2. If $x > y$ and $y > z$, then $x > z$.
$x > y$
$y > z$
$\therefore x > z$

3. If a $\parallel$ c and b $\parallel$ c, a $\parallel$ b.
a $\parallel$ c; b $\parallel$ c
$\therefore$ a $\parallel$ b

4. If a $\parallel$ b, $\angle x = \angle y$
a $\parallel$ b
$\therefore \angle x = \angle y$

5. If $x = y$ and $y = 3$, $x = 3$.
$x = y$
$y = 3$
$\therefore x = 3$

6. MN = AB/2 and MN $\parallel$ AB
$\therefore$ MN = AB/2
$\therefore$ MN $\parallel$ AB

7. A point P is called the *pole* of a line p if and only if P is not joined to any point on p by a line.
P is joined to a point on p by a line.
$\therefore$ P is not the pole of p.

8. Angle ACB is an inscribed angle in circle O.
If angle ACB is inscribed in circle O, it is measured by one-half of its intercepted arc AB.
$\therefore$ Angle ACB is measured by one-half of arc AB.

9. Triangles ABC and A′B′C′ are right triangles with hypotenuses c and c′, respectively, c = c′, $\angle$ABC = $\angle$A′B′C′.
$\therefore \triangle ABC \cong \triangle A'B'C'$.

10. Point X is on the perpendicular bisector of line segment AB if and only if AX = BX.
AX $\neq$ BX.
$\therefore$ X is not on the perpendicular bisector of line segment AB.

11. Consider the following conventional proof of the statement: The opposite sides of a parallelogram are equal.

Given: Quadrilateral ABCD is a parallelogram.
To Prove: AB = DC and AD = BC.

Statement	*Reason*
1. Quadrilateral ABCD is a parallelogram.	1. Given.
2. Draw diagonal AC.	2. There exists one and only one line through points A and C.
3. AB ∥ DC and AD ∥ BC.	3. Quadrilateral ABCD is a parallelogram if and only if its opposite sides are parallel.
4. ∠BCA = ∠DAC.	4. If AD ∥ BC and transversal AC falls on them, then alternate interior angles BCA and DAC are equal.
5. ∠BAC = ∠DCA.	5. Similarly.
6. AC = AC.	6. Identity.
7. △ABC ≅ △ADC.	7. Triangles ABC and ADC are congruent if ∠BCA = ∠DAC, ∠BAC = ∠DCA, and AC = AC.
8. AB = DC and AD = BC.	8. Triangles ABC and ADC are congruent if and only if their corresponding parts are equal.

Rewrite the proof so that the uses of inference schemes are explicit. Pay particular attention to the tacit uses of modus ponens and conjunctive inference.

You might have to supply some statements and steps that are omitted in this conventional proof. In addition, you might wish to recast the statements of some of the theorems and axioms used.

2.15 COLLECTION OF INFERENCE RULES

Important inference rules or schemes are collected below.

	Inference rule	*Name*	*Reference*
(2.71)	$\dfrac{P \to Q, P}{Q}$	Modus ponens	(2.64)
(2.72)	$\dfrac{P, Q}{PQ}$	Conjunctive inference	(2.65)
(2.73)	$\dfrac{PQ}{P}$	Conjunctive simplification	(2.66)
(2.74)	$\dfrac{PQ}{Q}$	Conjunctive simplification	(2.66)

(2.75)	$\dfrac{P \rightarrow Q,\ Q \rightarrow R}{P \rightarrow R}$	Hypothetical syllogism	(2.67)
(2.76)	$\dfrac{P \rightarrow Q}{\sim Q \rightarrow \sim P}$		(2.69)
(2.77)	$\dfrac{P \rightarrow Q,\ \sim Q}{\sim P}$	Contrapositive inference (also called *modus tollens*)	(2.70)
(2.78)	$\dfrac{A \rightarrow Q,\ B \rightarrow Q}{A \vee B \rightarrow Q}$	Inference by cases	

III: PROOF AND DEMONSTRATION

3.1 BASIC FORM OF INDIRECT PROOF

In an indirect proof of a statement "A", it is customary to show that the assumption of "$\sim A$" leads to a contradiction, so that "$\sim A$" must be false and thus "A" is true. A contradiction is understood to be the denial of an axiom or of a previously proved theorem. In addition, any statement known to be false because of its form alone can serve as a contradiction. Perhaps the simplest such form is "$R\sim R$", for whatever the truth value taken for "R", the truth value of "$R\sim R$" will clearly be "F". Therefore, any statement in the form "$R\sim R$" is false, and can serve as the contradiction in an indirect argument.

We shall take as the basic form of indirect proof of "A" any argument that starts from "$\sim A$" and leads to a statement in the form "$R\sim R$". This basic form includes the case where the contradiction is the denial of an axiom or theorem, for if "R" is an axiom say, and the argument leads to "$\sim R$", then by conjunctive inference (2.72) it also leads to "$R\sim R$".

The basic form of indirect proof is justified by the rule of contrapositive inference (2.77). The argument will establish "$\sim A \rightarrow R{\sim}R$". Since "$R{\sim}R$" is false, "$\sim(R{\sim}R)$" is true; so, by contrapositive inference,

$$\frac{\sim A \rightarrow R{\sim}R,\ \sim(R{\sim}R)}{A}$$

and "A" is proved.

In carrying out an indirect proof, one must be careful to start the argument with the negation of the statement to be proved, or with some statement equivalent to it. In most cases, the statement to be proved is a conditional of the form "$P \rightarrow Q$", so that the argument must start with "$\sim(P \rightarrow Q)$", which is grammatically awkward. Formula (2.41) provides a way out of the grammatical difficulty. Starting with "$\sim(P \rightarrow Q)$ as minor premise, and "$\sim(P \rightarrow Q) \leftrightarrow P{\sim}Q$" as major premise, "$P{\sim}Q$" can be inferred by modus ponens. In abbreviated form the inference is described

$$\frac{\sim(P \rightarrow Q) \leftrightarrow P{\sim}Q,\ \sim(P \rightarrow Q)}{P{\sim}Q}. \tag{3.1}$$

The indirect argument is completed by showing that "$P{\sim}Q$" leads to some contradiction "$R{\sim}R$".

In practice, the use of formula (2.41) and modus ponens is usually tacit, so that the argument appears to start simply from the single statement "$P{\sim}Q$". If we like, we can suppose the argument to start with the two statements "P" and "$\sim Q$", since each can be inferred by the inference schemes (conjunctive simplification) (2.73) and (2.74).

As an illustration, suppose that a, b, c are distinct lines in the plane, and we wish to prove the statement

A: If a $\|$ c and b $\|$ c, then a $\|$ b.

Since "A" is in the form of a conditional, we start the indirect proof with

(1) $\sim(P \rightarrow Q)$: Not; if a $\|$ c and b $\|$ c, then a $\|$ b.

By (3.1),

(2) $P{\sim}Q$: a $\|$ c and b $\|$ c, and a is not parallel to b.

Then, by the inference scheme (conjunctive simplification) (2.73),

(3) P: a $\|$ c and b $\|$ c;

and in the same way, using (2.74),

(4) $\sim Q$: a is not parallel to b.

From the definition of parallel lines,

(5) $\sim Q \rightarrow S$: If a is not parallel to b, then a and b meet at some point N.

From Steps 4 and 5 by modus ponens,

(6) S: a and b meet at N.

From Steps 3 and 6 by conjunctive inference

(7) PS: a $\parallel$ c and b $\parallel$ c, and a and b meet at N.

From an axiom of geometry we know that there is only one line through a given point parallel to a given line; so we have

(8) $\sim(PS)$: Not; a $\parallel$ c and b $\parallel$ c, and a and b meet at N.

So by conjunctive inference we have reached the contradiction

$$PS \sim (PS)$$

so that "$\sim(P \rightarrow Q)$", or "$\sim A$", is false, and thus "A" is true.

The conventional language of indirect proofs can be quite misleading. In proofs of the foregoing theorem on parallels, the first statement of the indirect proof is frequently taken to be: "Suppose that a is not parallel to b". From this it would appear that the indirect proof of "$P \rightarrow Q$" starts with "$\sim Q$". Of course, if it were possible to show that "$\sim Q$" leads to a contradiction without using the hypothesis "P", then "Q" would be proved true. Then if "Q" is true, so is "$P \rightarrow Q$" true, as can be seen by consulting the truth table for the conditional (2.3). However, in our proof of the theorem, we used "P" as well as "$\sim Q$" to arrive at the contradiction "$PS \sim (PS)$". It is safe to say that if a theorem is worth stating, then it will not be possible to deduce a contradiction from the denial of its conclusion alone.

Again, the language "suppose a is not parallel to b" may appear to indicate that it is "$P \rightarrow \sim Q$" that is to be proved false. Now, "$(P \rightarrow \sim Q) \leftrightarrow \sim(P \rightarrow Q)$" is *not* a valid statement formula, so that this procedure is immediately suspect. Suppose it can be shown that "$P \rightarrow \sim Q$" leads to a contradiction. Then "$\sim(P \rightarrow \sim Q)$" is proved. This is an awkward form, but it turns out to be equivalent to "PQ". The equivalence follows from formula (2.41) by replacing "Q" by "$\sim Q$" and then substituting "Q" for "$\sim\sim Q$" to obtain

$$\sim(P \rightarrow \sim Q) \leftrightarrow PQ.$$

With this formula and modus ponens, "PQ" is proved. Now it is "$P \rightarrow Q$" that is supposed to be proved, and instead we have proved

"PQ". To see what has been accomplished by this, it is necessary to look at the relation between these two formulas. From the truth table,

P	Q	PQ	$P \rightarrow Q$
T	T	T	T
F	T	F	T
T	F	F	F
F	F	F	T

it is easy to see that the following formula is valid:

$$PQ \rightarrow (P \rightarrow Q). \tag{3.2}$$

[This is formula (2.60).] With "PQ" established, then with (3.2) by modus ponens, "$P \rightarrow Q$" can be inferred, and the theorem is proved. The difficulty with this method is that it requires us to prove too much. It can happen that "$P \rightarrow Q$" is true when "PQ" is false, and in this case, which is the usual one, the method of proof fails. We say that this is the usual case, because if the stronger statement "PQ" is provable, one would not ordinarily assert the weaker statement "$P \rightarrow Q$" as a theorem.

There is a correct way of proving a theorem in the form "$P \rightarrow Q$" that starts from "$\sim Q$". The method consists in starting from "$\sim Q$" and deducing "$\sim P$". Then "$\sim Q \rightarrow \sim P$" is proved and the theorem follows from formula (2.42) by modus ponens. This contrapositive form of proof is neater, and conceptually simpler, than an indirect proof leading to a contradiction, and should be preferred when both methods are available.

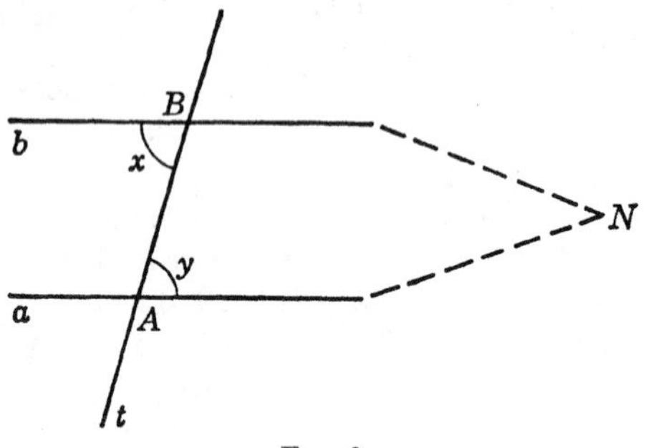

Fig. 6.

The two methods are compared in the following proofs of a standard geometry theorem on parallels cut by a transversal. Suppose lines a and b are cut by a transversal t as in Fig. 6, and it is desired to prove

$$P \rightarrow Q: (\angle x = \angle y) \rightarrow (a \parallel b).$$

An indirect proof leading to a contradiction might start with

(1) $P{\sim}Q$: ($\angle$x = $\angle$y)&(a is not parallel to b).

Then by the inference scheme conj simp (2.73),

(2) P: $\angle$x = $\angle$y,

and in the same way (2.74),

(3) ${\sim}Q$: a is not parallel to b.

From the definition of parallel lines,

(4) ${\sim}Q \rightarrow S$: (a is not parallel to b) $\rightarrow$ (a and b meet at a point, say N).

By modus ponens, from Steps 3 and 4,

(5) S: a and b meet at N.

Then ABN is a triangle, and we have to suppose that it has already been proved that "an exterior angle of a triangle is greater than either opposite interior angle" so that

(6) ${\sim}P$: $\angle$x $\neq$ $\angle$y.

Finally, from Step 2 and Step 6 by conjunctive inference,

(7) $P{\sim}P$: ($\angle$x = $\angle$y)($\angle$x $\neq$ $\angle$y),

and a contradiction has been reached.

This proof contains within it the contrapositive form of proof. Steps 3 to 6 give a deduction of "${\sim}P$" from "${\sim}Q$", that is, they prove "${\sim}Q \rightarrow {\sim}P$", and the theorem "$P \rightarrow Q$" follows from the contrapositive equivalence formula (2.42) and modus ponens.

3.2 SPECIAL CASES OF THE BASIC FORM OF INDIRECT PROOF

The basic form of indirect proof requires that the argument lead to a contradiction "$R{\sim}R$". In most cases, "R" is an axiom or a previously proved theorem, but occasionally it will be neither of these. It may sometimes be possible to deduce "Q" starting from "$P{\sim}Q$". But "${\sim}Q$" also follows from "$P{\sim}Q$" [see (2.74)]; so in this case the contradiction is "$Q{\sim}Q$" and we have just a special case of the basic form of indirect proof.

As an illustration of this special form of argument, let us prove the following statement in which x stands for some positive real number:

$P \rightarrow Q$: If $x \leq x^t$ for every positive number t less than 1, then $x \leq 1$.

We start the proof from

(1) $P{\sim}Q$: ($x \leq x^t$ for every positive number t less than 1)&($x > 1$).

From Step 1, by conj simp (2.73), it follows that

(2) P: $x \leq x^t$ for every positive number t less than 1,

and also, from (2.74),

(3) $\sim Q: x > 1.$

From Step 3 it follows that $1/x < 1$ so that we may use $1/x$ for t in Step 2 to obtain

(4) $x \leq x^{1/x}.$

Taking the logarithm of both members, we have

(5) $\log x \leq \log x^{1/x}$ or $\log x \leq (1/x) \log x.$

From Step 3 it follows that $\log x$ is positive, and so when we divide both sides of the inequality of Step 5 by $\log x$ we obtain

(6) $1 \leq 1/x,$

and from Step 6, it easily follows that

(7) $Q: x \leq 1.$

Finally, from Steps 3 and 7 by conjunctive inference,

(8) $Q{\sim}Q$: ($x \leq 1$)&($x > 1$),

and we have the expected contradiction.

It is worth noting that "$\sim Q$" plays an essential role in this proof in obtaining Step 4 and Step 6.

Sometimes it is possible to start with "$P{\sim}Q$" and deduce "$\sim P$". Since "P" is also deducible from "$P{\sim}Q$" (conj simp), the contradiction in this case is "$P{\sim}P$", and the argument is another special form of the basic indirect argument. This form of argument seems much like the contrapositive argument mentioned at the close of the preceding section. The difference is that in proving "$P \to Q$" by proving "$\sim Q \to \sim P$", it must be possible to find an argument that leads from "$\sim Q$" to "$\sim P$", which does not need "P" as a step, whereas in the argument leading from "$P{\sim}Q$" to "$\sim P$", it is necessary to use "P" as a step.

The hypotheses of theorems of geometry and algebra are very often conjunctions. If a theorem has the form "$ABC \to Q$", then it may be possible to show that "$ABC{\sim}Q$" leads to "$\sim A$" (or "$\sim B$" or "$\sim C$"). The required contradiction will then be "$A{\sim}A$" (or "$B{\sim}B$" or "$C{\sim}C$", respectively).

As an example of proving a theorem of the form "$AB \to Q$" by this method, we prove: "The sum of a rational number and an irrational number is irrational". That is,

$$AB \rightarrow Q\text{: } (x \text{ rational})(y \text{ irrational}) \rightarrow (x + y \text{ irrational}).$$

We start the proof by assuming

$$(1) \qquad AB{\sim}Q\text{: } (x \text{ rational})(y \text{ irrational}){\sim}(x + y \text{ irrational}).$$

As in previous illustrations, we have the steps:

(2) A: x rational (by conj simp)
(3) B: y irrational (by conj simp)
(4) ${\sim}Q$: ${\sim}(x + y$ irrational),

or what is the same,

$$x + y \text{ rational (by conj simp).}$$

Now,

$$(x + y) - x = y$$

is true because it is an instance of an algebraic identity. Furthermore, it is known that the difference of two rational numbers is rational; so by Steps 2 and 4, $(x + y) - x$ is rational, and thus y is rational. That is,

$$(5) \qquad {\sim}B\text{: } {\sim}(y \text{ irrational}).$$

By conjunctive inference (Steps 3 and 5) we have "$B{\sim}B$"; hence, a contradiction, and the theorem is true.

The form of proof that we have called the basic form of indirect proof is used a great deal in elementary mathematics. In order that the notion of this form of indirect proof be clear, we summarize the format for such proof as follows:

a. The basic form of the indirect proof of a statement "A" is an argument that starts from "${\sim}A$" and leads to a contradiction "$R{\sim}R$".
b. If "A" is a conditional "$P \rightarrow Q$", then the argument can start from "$P{\sim}Q$".
c. If "A" is a conditional "$P \rightarrow Q$", a contradiction can be reached if the argument leads to:
   1. The denial of an axiom
   2. The denial of a previously proved theorem
   3. "Q"
   4. "${\sim}P$"

3.3 PROOF BY ELIMINATION

The so-called "proof by the process of elimination" can be the cause of considerable bewilderment to the beginning mathematics student. The method is essentially indirect, and indeed is often presented in elementary geometry textbooks as the fundamental pattern for indirect proofs.

The following recipe for carrying out such proofs, while not a direct quotation, is a fair paraphrasing of such recipes as they occur in several current geometry textbooks.

To prove a theorem by the process of elimination:

a. State the conclusion and *all* the other possibilities.
b. Show that the other possibilities each leads to a contradiction of the hypothesis or of a truth previously learned.
c. The conclusion is the only remaining possibility.

This apparently explicit set of directions is really quite loose and incomplete. Suppose that the theorem to be proved is a statement of the form "$P \rightarrow A$". For the moment, assume that the first statement of the recipe is understood and that there is just one other possibility "B". Following the second directive of the recipe, one goes on to show that "B" leads to some contradiction; hence "B" is false and "$\sim B$" is true. Then according to the third directive, "A" is the only other possibility. However, the goal is to prove "$P \rightarrow A$", and it is not yet clear how this has been done. To clear up this point, it is necessary to investigate what is meant by the phrase " . . . all the other possibilities" in the first statement of the recipe.

The phrase " . . . all the other possibilities" can be interpreted as meaning a set of alternatives to "A", that is, a set of additional statements "B", "C", . . . , "F", such that "$P \rightarrow A \vee B \vee C \vee \ldots \vee F$" is known to be true. With this interpretation, to say that there is just one alternative to "A" is to say that "$P \rightarrow A \vee B$" is known to be true. Let us examine this case of just one alternative "B" to "A". Now, "$\sim B$" is established by showing that "B" leads to some contradiction. If we interpret the third statement of the recipe to mean that now "$P \rightarrow A$" is proved, we are asserting that the inference scheme

$$\frac{P \rightarrow A \vee B,\ \sim B}{P \rightarrow A}$$

is correct. It is correct, by mod pon and (2.52), but is not a very practical way of eliminating "B", because it is hardly ever possible to reach a contradiction from "B" alone without making use of the hypothesis "P". But if "P" is used, the elimination argument actually starts from "PB", so that if a contradiction is reached, it is "$\sim(PB)$" which is established. The inference scheme that must be verified in this case is

$$\frac{P \rightarrow A \vee B,\ \sim(PB)}{P \rightarrow A}. \tag{3.3}$$

The inference scheme (3.3) can be derived from the valid formulas (2.58) and (2.53), which we restate for convenience:

$$\sim(PB) \leftrightarrow (P \to \sim B)$$
$$(P \to A \vee B) \leftrightarrow [(P \to \sim B) \to (P \to A)].$$

By substituting the first of these into the second, we obtain

$$(3.4) \qquad (P \to A \vee B) \leftrightarrow [\sim(PB) \to (P \to A)].$$

Now the inference scheme (3.3) can be verified, for from "$P \to A \vee B$" and (3.4) we get

$$(3.5) \qquad \sim(PB) \to (P \to A)$$

by modus ponens. Then from "$\sim(PB)$" and (3.5) by modus ponens we get

$$P \to A.$$

In some cases it will be possible to eliminate "B" by showing that it leads to "$\sim P$", thus proving "$B \to \sim P$". Since "$(B \to \sim P) \leftrightarrow (P \to \sim B)$" is an instance of contrapositive equivalence, and since "$\sim(PB) \leftrightarrow (P \to \sim B)$" has already been proved valid, it is easy to see that this type of elimination is also justified by the inference scheme (3.3).

To summarize, if "$P \to A \vee B$" is known to be true, then if "PB" leads to a contradiction (which may be "$P \sim P$"), or if "B" leads to "$\sim P$", the proof of "$P \to A$" is established. We take this as the basic form of elimination argument. To generalize to the case where there is more than one alternative to "A" is not difficult. The case of just two alternatives should be sufficient for illustration.

Suppose that "$P \to A \vee C \vee D$" is known to be true. Then "$P \to A$" will be proved if it can be shown that both "PC" and "PD" lead to contradictions. For then, "$\sim(PC)$" and "$\sim(PD)$" are established so that, by conjunctive inference, we have "$\sim(PC)\sim(PD)$". Now, by replacement in the De Morgan formula (2.39),

$$\sim(PC)\sim(PD) \leftrightarrow \sim(PC \vee PD),$$

and by substitution using the distributive law (2.48),

$$\sim(PC \vee PD) \leftrightarrow \sim[P(C \vee D)].$$

Then, by substitution,

$$\sim(PC)\sim(PD) \leftrightarrow \sim[P(C \vee D)].$$

To justify the outlined elimination procedure, we must verify the inference scheme

$$(3.6) \qquad \frac{P \to A \vee (C \vee D),\ \sim[P(C \vee D)]}{P \to A}.$$

But (3.6) is just (3.3) with the placeholder "B" replaced by "$C \vee D$", so the justification is completed.

As an illustration of a proof by elimination of two alternatives, consider the familiar plane geometry theorem:

> In a given circle, if chord AB is greater than chord CD, then chord AB is closer to the center than chord CD.

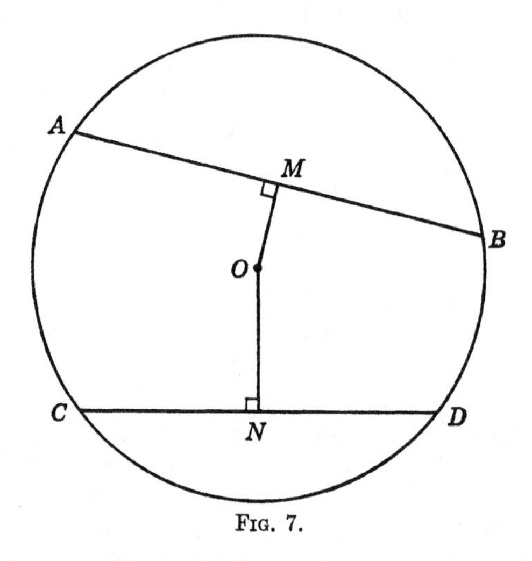

FIG. 7.

In terms of Fig. 7 as lettered, the statement to be proved may be written

$$P \rightarrow A: (\mathrm{AB} > \mathrm{CD}) \rightarrow (\mathrm{OM} < \mathrm{ON}).$$

Here the disjunction of alternatives may be taken as

$$(3.7) \qquad A \vee B \vee C: (\mathrm{OM} < \mathrm{ON}) \vee (\mathrm{OM} = \mathrm{ON}) \vee (\mathrm{OM} > \mathrm{ON}).$$

Observe that in (3.7),

$$\sim A \leftrightarrow B \vee C,$$

that is,

$$\mathrm{OM} \not< \mathrm{ON} \leftrightarrow (\mathrm{OM} = \mathrm{ON}) \vee (\mathrm{OM} > \mathrm{ON}).$$

Thus, (3.7) can be written as "$A \vee \sim A$", which is in the form of a valid formula and is therefore a true statement. Since (3.7) is true, then by the definition of the conditional,

$$P \rightarrow A \vee B \vee C: (\mathrm{AB} > \mathrm{CD}) \rightarrow (\mathrm{OM} < \mathrm{ON}) \vee (\mathrm{OM} = \mathrm{ON}) \vee (\mathrm{OM} > \mathrm{ON})$$

is true. The proof may be completed by eliminating the two alternatives as outlined in the previous discussion. Briefly, "$PB \rightarrow \sim P$" because by a previously proved theorem, "$(\mathrm{OM} = \mathrm{ON}) \rightarrow (\mathrm{AB} = \mathrm{CD})$".

The contradiction in this case will be "$P{\sim}P$". Also, "$PC \to {\sim}P$", because we suppose it has already been proved that "(OM > ON) $\to$ (AB < CD)". Again the contradiction is "$P{\sim}P$". We could also have achieved the eliminations here by proving "$B \to {\sim}P$" and "$C \to {\sim}P$".

The reasoning that assured us of the truth of "$P \to A \vee B \vee C$" in the foregoing illustration is typical of elimination proofs. In the general case of several alternatives, "$P \to A \vee B \vee \cdots \vee F$" is most often known to be true because "$A \vee B \vee \cdots \vee F$" has the special form "$A \vee {\sim}A$", that is,

$$B \vee \cdots \vee F \leftrightarrow {\sim}A.$$

Commonly, "A" and its various alternatives have the property that no two of them can both be true. This is the case in (3.7) of the illustration, but such a property is not a requirement for a proof in this form. For instance, it is known of the decimal representation of any real number x that

(3.8) (dec rep of x terminates)v(dec rep of x is periodic)v(x is irrational).

The first two statements in (3.8) are not exclusive, for 0.2 and 0.1999 . . . are decimal representations of the same rational number 2/10, although the first terminates and the second is periodic. If it is desired to prove some real number is irrational, given some facts about its decimal representation, then (3.8) could serve as the disjunction of alternatives in a proof by elimination.

The recipe for proof by elimination, mentioned at the beginning of this section, can be restated with more precision as follows:

To prove "$P \to A$" indirectly by elimination:

a. State enough alternatives "B", "C", . . . , "F" to "A" so that either "$P \to A \vee B \vee C \vee \cdots \vee F$" or "$A \vee B \vee C \vee \cdots \vee F$" is known to be true.

b. Eliminate "F" by showing that "$F \to {\sim}P$" or that "PF" leads to a contradiction "$R{\sim}R$".

c. If all the other alternatives to "A" are eliminated similarly, then "$P \to A$" is proved.

We have analyzed this type of argument at some length because it is so often treated in elementary textbooks as the general pattern for all indirect proofs. However, it is not so simple as the form we have presented as the basic form of indirect proof, and it is surely harder to describe. If "$P \to Q$" can be proved by showing that "$P{\sim}Q$" leads to a contradiction, or by proving the contrapositive "${\sim}Q \to {\sim}P$", it seems ridiculous to go through the ritual of saying, "Either 'Q' is true, or it is false", in order to cast the argument into the more complex form of a proof by elimination.

EXERCISES

1. Prove: If two chords in a circle are unequal, then their arcs are unequal.

2. Prove: If two sides of triangle ABC are unequal, then the angles opposite these sides are unequal.

3. Prove: If a point is not on the bisector of an angle, then it is not equidistant from the sides of the angle.

4. Prove: $\sqrt{2}$ is not rational.

5. Prove: $\sqrt{3}$ is not rational.

6. Prove: The product of a nonzero rational number and an irrational number is irrational.

7. Prove: There is no largest prime number.

8. Prove: If x is a positive real number less than 1 such that in its decimal representation, the nth decimal place is occupied by the digit 1 when n is a power of 2, and otherwise the nth decimal place is occupied by a digit different from 1, then x is irrational. Use (3.8).

9. Read the presentations of indirect proof in two different plane geometry textbooks and analyze the descriptions in view of the discussions in Sec. 3.2 and 3.3.

10. Analyze three examples of the use of indirect proof in elementary mathematics.

3.4 PROOF BY CASES

In proving the plane geometry theorem:

> An angle inscribed in a circle is equal in degrees to one-half of its intercepted arc,

it is customary to break the proof up into three cases, corresponding to the three diagrams of Fig. 8. To analyze the proof by cases formally, let us take as translations:

P: x is inscribed in circle O.
A: x inscribed in O has one side through the center.
B: x inscribed in O has its sides on opposite sides of the center.
C: x inscribed in O has both its sides on the same side of the center.
Q: x is equal in degrees to one-half of its intercepted arc on circle O.

With these translations, the theorem to be proved is "$P \rightarrow Q$". Its proof is established by proving the three cases:

Case I: $A \rightarrow Q$; Case II: $B \rightarrow Q$; Case III: $C \rightarrow Q$.

Informally, the idea here is, given "P", then at least one of the statements "A" or "B" or "C" is true, and since each of them implies "Q"

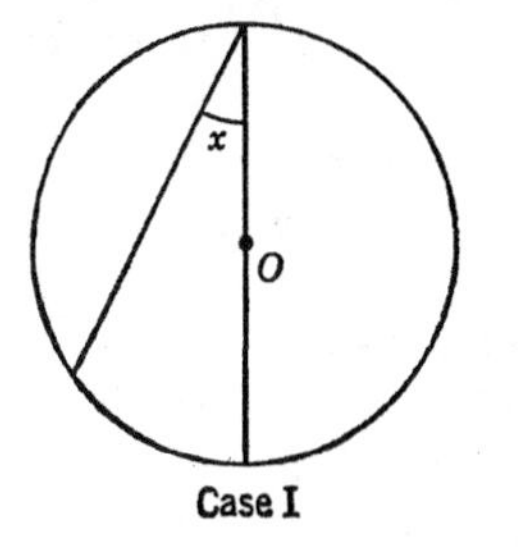

Case I

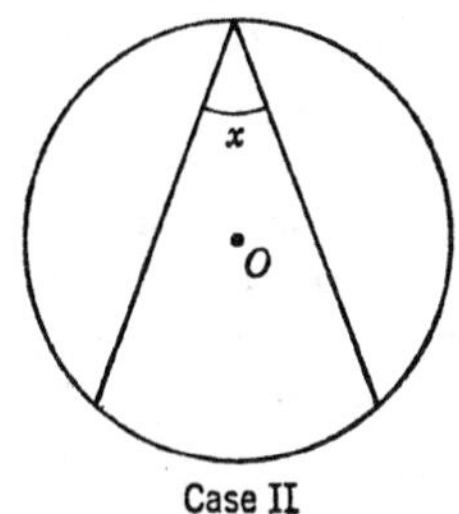

Case II

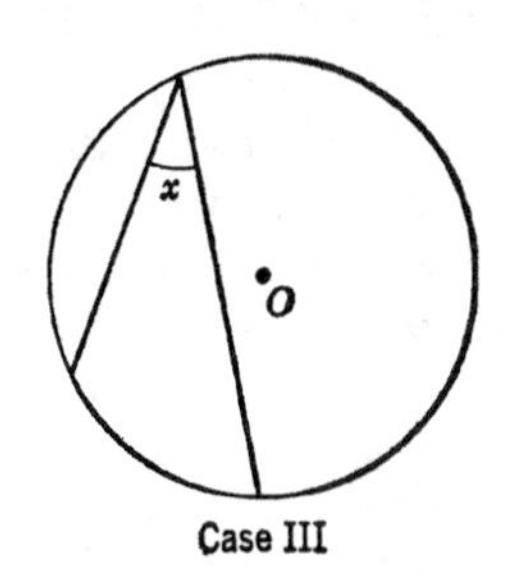

Case III

FIG. 8.

then "$P \rightarrow Q$". The validity of the procedure depends on the correctness of two steps. First, it is necessary to describe the cases so that we know

$$P \rightarrow A \vee B \vee C,$$

and second, the inference scheme,

$$\frac{A \rightarrow Q,\ B \rightarrow Q,\ C \rightarrow Q}{A \vee B \vee C \rightarrow Q}, \tag{3.9}$$

must be verifiable. To verify the inference scheme, we start with

$$A \rightarrow Q, \qquad B \rightarrow Q, \qquad C \rightarrow Q$$

and infer

$$(A \rightarrow Q)(B \rightarrow Q)(C \rightarrow Q) \tag{3.10}$$

by successive applications of conjunctive inference. Next, we need to know that the following formula is valid:

$$(A \rightarrow Q)(B \rightarrow Q)(C \rightarrow Q) \leftrightarrow (A \vee B \vee C \rightarrow Q). \tag{3.11}$$

Formula (3.11) can be obtained from (2.54) by substitution and replacement, or may be proved valid by truth tables. Then, from (3.10) and (3.11), by modus ponens, we infer

$$A \vee B \vee C \rightarrow Q.$$

The inference scheme (3.9) summarizes this sequence of inferences.

Once "$P \to A\vee B\vee C$" and "$A\vee B\vee C \to Q$" have been established, "$P \to Q$" can be inferred by the hypothetical syllogism (2.75).

It usually happens in a proof by cases that the hypothesis "P" of the theorem to be proved not only implies, but is equivalent to, the disjunction of the hypotheses in the cases. This is the situation in the foregoing illustration where for "$P \to A\vee B\vee C$" we could have taken "$P \leftrightarrow A\vee B\vee C$". It is also frequently the case that the disjunction of hypotheses is exclusive, but this is not necessary to the validity of the outlined procedure.

Little is said about the method of proof by cases in most geometry textbooks. It is not used with very many theorems, and when it does occur, it is generally accepted by students as a natural way of proving theorems. The method might well be used more freely in geometry. Its use, in conjunction with contrapositive forms of proof, and the basic indirect form, can replace all indirect proofs by elimination.

Consider how the illustration of Sec. 3.3 can be handled as a proof by cases. Recall that in terms of the notation of Fig. 7, the theorem to be proved is

$$P \to A\colon (\mathrm{AB} > \mathrm{CD}) \to (\mathrm{OM} < \mathrm{ON}).$$

As before, it is assumed known from previous theorems that

$$(\mathrm{OM} = \mathrm{ON}) \to (\mathrm{AB} = \mathrm{CD}) \tag{3.12}$$

and

$$(\mathrm{OM} > \mathrm{ON}) \to (\mathrm{AB} < \mathrm{CD}). \tag{3.13}$$

The theorem can be proved by proving its contrapositive

$$\sim A \to \sim P\colon (\mathrm{OM} \not< \mathrm{ON}) \to (\mathrm{AB} \not> \mathrm{CD}).$$

Since "$\mathrm{OM} \not< \mathrm{ON}$" is the same as "$(\mathrm{OM} = \mathrm{ON})\vee(\mathrm{OM} > \mathrm{ON})$", the contrapositive can be proved by the cases:

Case I: $(\mathrm{OM} = \mathrm{ON}) \to (\mathrm{AB} \not> \mathrm{CD})$. [This follows easily from (3.12).]
Case II: $(\mathrm{OM} > \mathrm{ON}) \to (\mathrm{AB} \not> \mathrm{CD})$. [This follows easily from (3.13).]

The form of proof by using the contrapositive and cases seems to us simpler in conception than the proof of the same theorem by the method of elimination of Sec. 3.3.

It is worth noticing that in a proof by cases, there is no requirement that all the cases be proved with the same method. In the foregoing proof, a direct proof is simplest for both cases, but for other theorems it may turn out to be simpler if some of the cases are proved directly and some indirectly.

In summary, where an indirect proof seems indicated there are three main procedures to consider. These procedures are adequate for elementary mathematics, and appear to us to be the simplest possible. To prove a theorem in the form "$P \rightarrow Q$":

a. By a contrapositive method
Prove "$\sim Q \rightarrow \sim P$" instead. When this is possible, it seems the neatest of the three procedures.

b. By proving the contrapositive by cases
If in the contrapositive "$\sim Q \rightarrow \sim P$", "$\sim Q$" is easily expressed as a disjunction "$A \vee B \vee \ldots \vee F$", then prove the cases, "$A \rightarrow \sim P$", "$B \rightarrow \sim P$", . . . , "$F \rightarrow \sim P$". Remember that any legitimate method can be used to prove the cases. Some of the cases may be proved indirectly and some directly.

c. By the basic indirect form
Assume "$P \sim Q$" and deduce a contradiction "$R \sim R$". Remember that the contradiction can be "$P \sim P$" or "$Q \sim Q$".

EXERCISES

1. Assume it has been proved that:

> If two sides of a triangle are unequal, then the angle opposite the longer side is larger than the angle opposite the shorter side.

State a converse and prove it by cases.

2. Assume it has been proved that:

> If two triangles have two sides of one equal, respectively, to two sides of the other and the included angle of the first is greater than the included angle of the second, then the third side of the first is greater than the third side of the second.

State a converse, and prove it by cases.

3. Look up the proof of the law of cosines in some standard trigonometry textbook. Analyze the proof in view of the discussion in Sec. 3.4.

4. Repeat Exercise 3 for the proof of the formula for sin $(\alpha + \beta)$.

5. State a converse of the illustrative example of Sec. 3.3, and prove it by cases.

3.5 CONVERSES

In contrast to the notion of *conversion* in Aristotelian logic, the formal notion of *converse theorem*, as used in modern mathematics, is not clearly defined. The converse of "$P \rightarrow Q$", where "P" and "Q" are simple enough statements, is commonly taken to be "$Q \rightarrow P$". For complicated statements there is no general agreement as to the proper way to form the converse. The various ways of forming converses do have in common some sort of interchange of statements between the antecedent and consequent of a conditional.

Elementary mathematics textbooks vary in treatment of conversion. Some simply describe the converse of a theorem as formed by "interchanging the hypothesis and the conclusion". This is hardly adequate, since few theorems of geometry are actually converted in this way. Others discuss the notion that a theorem may have several related statements that are in some sense its converses. These are distinguished as complete converses and partial converses.

Statements that can be put in the form "$P \to (Q \to R)$" are common in elementary mathematics. Should the converse be taken as "$(Q \to R) \to P$", or as "$P \to (R \to Q)$", or perhaps as "$(R \to Q) \to P$"? Examples of at least the first two forms occur in both elementary and higher mathematics.

As an instance of the form "$P \to (Q \to R)$", let us state a theorem of plane geometry about a triangle ABC, where the letter m denotes a line.

(3.14) $P \to (Q \to R)$: If AC = BC, then m bisects ∠C implies that m is the perpendicular bisector of AB.

Possible converses of (3.14) might be taken as

(3.15) $(Q \to R) \to P$: If m bisects ∠C implies that m is the perpendicular bisector of AB, then AC = BC,

(3.16) $P \to (R \to Q)$: If AC = BC, then m the perpendicular bisector of AB implies that m bisects ∠C,

(3.17) $(R \to Q) \to P$: If m is the perpendicular bisector of AB implies that m bisects ∠C, then AC = BC.

In an attempt to select one of these as the proper converse, one might argue that in (3.14) the implication symbol immediately following "P" indicates the *main* implication so that the proper choice for the converse should be (3.15). If there is a way of locating the *main* implication in a statement, then conversion of the statement might be carried out by interchanging the antecedent and consequent of the main implication. Unfortunately, to define the operation of conversion in this way is not quite satisfactory, for it can happen that two equivalent statements have non-equivalent converses. For instance, the validity of the following formula (2.55) was proved as an exercise in Sec. 2.9.

$$[P \to (Q \to R)] \leftrightarrow [PQ \to R].$$

But the formula

$$[(Q \to R) \to P] \leftrightarrow [R \to PQ]$$

is not valid. (Take for "P", "Q", "R" the values "T", "F", "T", respectively.)[1]

Conversion may be thought of as an operation that yields new statements from given conditional statements. As currently used, the operation certainly need not yield true statements when applied to true statements, and when the operation is applied to equivalent statements, it need not even yield equivalent statements. An operation having such properties is not a very useful tool in formal logic.

To gain deeper insight into the meaning of a theorem, it is valuable to consider formally related statements that may be proved, or disproved, as theorems. The statements (3.15) to (3.17) are all conditionals formally related to (3.14) in the sense that they are formed out of substatements contained in (3.14). Proof or disproof of any of them will surely shed light on the properties of isosceles triangles. Perhaps it is this activity of investigating statements related to a proved theorem that is important in elementary mathematics, and this may be carried on without attempting a precise formal definition of conversion. The phrase "a converse" might be used to indicate any statement derived from a theorem by interchanging some statements in its hypothesis with some statements in its conclusion. Certainly the phrase "the converse" is ambiguous as used in most elementary textbooks.

3.6 CONVERSION IN CLASSICAL LOGIC

A basic deductive unit of Aristotelian logic is the *syllogism*, which was discussed briefly in Sec. 1.2. Recall that the syllogism is a form of argument dealing with four kinds of quantified statements called the A, E, I, O propositions, as in the following examples:

A. All soldiers are volunteers.
E. No soldiers are volunteers.
I. Some soldiers are volunteers.
O. Some soldiers are not volunteers.

These four quantified statements express various degrees of relationship between the classes "soldiers" and "volunteers". We list them again with the variables "s" and "p" representing placeholders for class names:

A. All s is p.
E. No s is p.
I. Some s is p.
O. Some s is not p.

[1] It is worth noting that neither of the other forms, (3.16) or (3.17), is equivalent to "$R \rightarrow PQ$".

There is a large, but finite, number of forms of the syllogism that can be constructed using various combinations of the categorical propositions. In classical logic, precise rules are given for determining the 19 forms representing valid forms of reasoning.

Certain immediate forms of reasoning allow conclusions to be drawn from a single proposition as premise. It is clear that *A* and *O* propositions with the same class names are opposite in truth value, so that if we know, for instance, that

All soldiers are volunteers

is a true (false) statement, then we know that

Some soldiers are not volunteers

is a false (true) statement. In the same way, the *E* and *I* propositions are opposite in truth value, so that the truth of one determines the truth of the other.

Other immediate forms of reasoning were supposed to exist in classical logic. For instance, it was held that the *A* proposition implies the *I* proposition having the same class names. So that from

All pigeons are birds

could be inferred

Some pigeons are birds.

The operation of conversion was defined in such a way that the converse of a true proposition was supposed to be true. In the case of an *E* or an *I* proposition, the converse was formed by simply interchanging the class names. Thus the *E* propositions

No American is a traitor.
No traitor is an American.

would be considered mutually converse, as would be the *I* propositions

Some algebraic numbers are irrational numbers.
Some irrational numbers are algebraic numbers.

In the case of a true *A* proposition, simple interchange of the class names will not necessarily yield a true proposition. In this case, conversion required a change of quantification as well as an interchange of class names. In classical conversion, the *A* proposition

All pigeons are birds

becomes the *I* proposition

Some birds are pigeons.

This type of conversion was called *conversion per accidens*, and it was supposed that the A proposition implied its I converse. Conversion was not allowed in the case of the O proposition.

We may summarize the Aristotelian conversion rules in the following table:

Proposition	*Converse*
A. All s is p.	I. Some p is s.
E. No s is p.	E. No p is s.
I. Some s is p.	I. Some p is s.
O. Some s is not p.	No converse.

It was supposed that with these rules of conversion, a proposition implies its converse. This property of the operation makes it a useful tool in classical logic.

However, it must be observed that it is a tacit assumption of Aristotelian logic that every class mentioned in a categorical proposition is nonempty. Thus the proposition

All unicorns are four-legged

is not allowed if the class of unicorns is considered to be empty. Or put another way, if the proposition is asserted, it is understood that the class of unicorns (and therefore also the class of four-legged things) has at least one member. For mathematics it is found desirable to use a logic that allows statements about empty classes. It is desirable, so to speak, that the logic take empty classes in its stride. The modern tendency is to take the universal A and E propositions as statements having no existential import and to take the particular I and O propositions as assertions of existence. Thus the statement

All unicorns are four-legged

is allowed, and it is understood that nothing has been asserted about the existence or nonexistence of unicorns or of four-legged things. However, if the class "unicorns" is known to be empty, the statement is considered to be true. On the other hand, the statement

Some unicorns are four-legged

is taken as an assertion of existence, and is considered false if there are no unicorns at all, or if there are no four-legged things.

The statement

All unicorns are four-legged

may be paraphrased as a conditional,

For any thing, if it is a unicorn,
then it is four-legged.

In the paraphrase, the two occurrences of the pronoun "it" clearly refer to the word "thing". We may rewrite the paraphrase

(3.18) For any x, if x is a unicorn, then x is four-legged.

Now if there are indeed no unicorns, then for every x, "x is a unicorn" is false. But then (3.18) is true in accordance with the definition of "$P \rightarrow Q$". Thus, taking the statement

All unicorns are four-legged

to be true when the class "unicorns" is empty is quite in accord with our notion of the conditional. *Conversion per accidens* of this statement yields

Some four-legged things are unicorns.

Then if this statement is taken to mean that there is at least one four-legged thing that is a unicorn, and if the class "unicorns" is indeed empty, the statement must be considered false. It follows that *conversion per accidens* does not always yield a true converse from a true statement when empty classes are permitted.

These difficulties can be avoided by ruling out of the logic all statements about empty classes, but particularly for mathematics it is much more desirable to alter the logic in such a way that empty classes can be treated on the same footing as nonempty classes. In such a logic, the Aristotelian notion of conversion has to be dropped. In modern mathematics, the term "converse" is generally applied only to conditionals, and although the term may be loosely used, mathematicians unanimously agree that the converse of a true conditional need not be true.

3.7 INVERSES

The term "inverse" or "inverse theorem" occurs in some plane geometry textbooks. By the inverse of a theorem in the form "$P \rightarrow Q$" is meant a conditional of the form "$\sim P \rightarrow \sim Q$". The word "inverse" is used in a variety of ways in higher mathematics, but virtually never in the above sense.

Considered as a formal operation, "inversion" is about as unsatisfactory as the operation "conversion". The inverse of a true conditional is not necessarily true (or necessarily false), and inverses of equivalent conditionals need not be equivalent. For example, as has been already verified,

$$[P \rightarrow (Q \rightarrow R)] \leftrightarrow [PQ \rightarrow R]$$

is a valid formula, but for the inverses,

$$[\sim P \rightarrow \sim(Q \rightarrow R)] \leftrightarrow [\sim(PQ) \rightarrow \sim R]$$

is not a valid formula, as may be verified by truth tables.

The relationship between a theorem and its contrapositive, converse, and inverse is frequently displayed in the diagram

$$\begin{array}{ccc} \text{Theorem} & \nwarrow\!\!\!\nearrow & \text{Converse} \\ \text{Inverse} & \swarrow\!\!\!\searrow & \text{Contrapositive} \end{array}$$

in which the double-headed arrows are supposed to indicate equivalence. For a theorem in the form "$P \rightarrow Q$", the diagram may be given

$$\begin{array}{ccc} P \rightarrow Q & \nwarrow\!\!\!\nearrow & Q \rightarrow P \\ \sim P \rightarrow \sim Q & \swarrow\!\!\!\searrow & \sim Q \rightarrow \sim P \end{array}$$

This type of diagram is apt to appear in the discussion of "locus" in plane geometry, where it is necessary to prove theorems of the form "$P \leftrightarrow Q$". It is pointed out that in order to prove equivalence, it is sufficient to prove any pair of statements in the diagram not diagonally opposite one another. Either diagram, and discussion of it, is designed to teach greater flexibility in handling theorems of the "if and only if" type, which always state an equivalence.

There is nothing about the diagrams and the rule for their use that would not already be known to a student who understands contrapositive equivalence. It seems more direct to point out that "$P \leftrightarrow Q$" can be proved by proving both "$P \rightarrow Q$" and "$Q \rightarrow P$", and to remind the student that it may be easier to prove "$P \rightarrow Q$" by proving its contrapositive "$\sim Q \rightarrow \sim P$" instead, and similarly, it may be easier to prove "$Q \rightarrow P$" by proving "$\sim P \rightarrow \sim Q$". Nothing would be gained here by introducing the notion of "inverse". Even less would be gained by using "inverse of the converse" in place of "contrapositive" as some authors do.

As an example, consider the standard locus theorem:

> The locus of a point X that is equidistant from two given points A and B is the perpendicular bisector m of the segment AB.

To prove this, it is necessary to prove the two statements:

(3.19) $P \rightarrow Q$: If AX = BX, then X is on m.

(3.20) $Q \rightarrow P$: If X is on m, then AX = BX.

If it is easier to prove (3.19) by proving its contrapositive, then one would prove

$\sim Q \rightarrow \sim P$: If X is not on m, then AX $\neq$ BX;

and if it is easier to prove (3.20) by proving its contrapositive, then one would prove

$\sim P \rightarrow \sim Q$: If AX $\neq$ BX, then X is not on m.

Finally, it might be that the basic indirect form of proof is the easiest way of proving (3.19) or (3.20).

EXERCISES

For each locus theorem (in the plane), write out statements in the forms "$P \rightarrow Q$" and "$Q \rightarrow P$" whose proofs will constitute a proof of the theorem. Write out the corresponding contrapositives, and decide which pair of the four statements you have written you would choose to prove. Consider the case for an indirect proof in Exercises 1 and 2.

1. The locus of a point equidistant from the sides of an angle is the bisector of the angle.

2. The locus of the vertex of an isosceles triangle with a given base is the perpendicular bisector of the base.

3. The locus of a point equidistant from two intersecting lines a and b is two lines that bisect the angles formed by a and b.

4. The locus of the vertex of the right angle of a right triangle having a given hypotenuse is a circle whose diameter is the given hypotenuse. (Strictly speaking, the end points of the hypotenuse are not part of the locus.)

5. The equation of the locus of a point at distance r from the origin is $x^2 + y^2 = r^2$.

6. The equation of the locus of a point such that the sum of its distances from $(-c,0)$ and $(c,0)$ is $2a$, where $c^2 = a^2 - b^2$ and $a > c$, is $x^2/a^2 + y^2/b^2 = 1$.

7. Verify that to prove a statement of the form "$P \rightarrow (Q \leftrightarrow R)$", it is sufficient to prove "$P \rightarrow (Q \rightarrow R)$" and "$P \rightarrow (R \rightarrow Q)$". Consider contrapositive variations of these statements.

8. Can you infer "$P \rightarrow (Q \leftrightarrow R)$" from "$PQ \rightarrow R$" and "$PR \rightarrow Q$"?

9. What statements would you prove to establish the theorem:

If ABC is a triangle, then AC = BC if and only if $\angle$A = $\angle$B.

10. What statements would you prove to establish the theorem:

If a and b are real numbers, then $ab = 0$ if and only if $a = 0$ or $b = 0$.

3.8 DEMONSTRATION AND PROOF

Up to this point, only short arguments have been considered. We have considered what might be called "units of deduction". Some abbreviated *demonstrations* have been presented, and *proof* has been discussed informally without any attempt at precise definitions of these notions.

In this section, a formal criterion is provided for accepting or rejecting any string of statements offered as proof of a theorem.

A theorem of plane geometry is a statement within the geometry (i.e., a statement about points, lines, planes, distance, etc.) that follows from the basic geometric axioms. To decide whether or not a statement is a theorem, one tries to construct a demonstration that the statement does indeed follow from the axioms, or tries to construct a demonstration that the denial of the statement follows from the axioms.

In attempting to discover statements that are likely candidates for theorems, every available means is employed, including considerations of meaning and form. Successful discovery is generally the result of experience, hard work, and luck. When a likely candidate for a theorem is discovered, construction of its demonstration is also aided by considerations of meaning. But the demonstration itself is the expression of a formal argument, and its validity depends solely on its form. The definition of a demonstration given in the next paragraph is couched in terms of form alone, and the task of checking a demonstration for validity is just a mechanical process of checking the form of each statement in it against this definition.

A *demonstration* that a statement "P" follows as a consequence of statements "$A_1, A_2, \ldots, A_n$" is a sequence of statements "$S_1, S_2, \ldots, S_k$" where "S_k" is the same as "P", and for each "S_i" in the sequence, one of the following is the case:

D_1: "S_i" is one of the statements "$A_1, A_2, \ldots, A_n$".
D_2: "S_i" is in the form of a valid statement formula.
D_3: "S_i" follows from earlier statements in the sequence by means of an established inference rule.
D_4: "S_i" is the same as some earlier statement in the sequence.

We express the notion that there is a demonstration that "P" follows from "$A_1, A_2, \ldots, A_n$" by writing

$$A_1, A_2, \ldots, A_n \vdash P. \tag{3.21}$$

The symbol "$\vdash$" is called a turnstile, and the statement (3.21) is usually read, "$A_1, \ldots, A_n$ yields P".[1]

The condition D_4 is included in the definition for convenience, but it is not essential. It permits repetition of statements in a demonstration, but for every demonstration of (3.21) having repetitions, there exists one not having repetitions. To see this, consider a demonstration, "$S_1, S_2, \ldots, S_k$" in which "S_j" is the same as an earlier statement "S_i". Then

[1] J. Barkley Rosser, *Logic for Mathematicians*, p. 56, New York: McGraw-Hill Book Company, Inc., 1953. Stephen Cole Kleene, *Introduction to Metamathematics*, p. 87, Princeton, N.J.: D. Van Nostrand Company, Inc., 1952.

if "S_j" is deleted from the sequence, the remaining $k - 1$ statements still satisfy the definition of a demonstration. In this manner, all repetitions may be eliminated to obtain a demonstration in which all the statements are distinct and satisfy the conditions D_1, D_2, D_3. Of course, it is a requirement that the last statement "S_k" be the statement we want to demonstrate. What happens if in the process of deleting redundancies, we delete "S_k"? Clearly, we may delete "S_k" only if it is the same as some earlier statement, say "S_r". But in this case, "$S_1, S_2, \ldots, S_r$" is already a demonstration, and the statements "$S_{r+1}, \ldots, S_k$" are redundant in the demonstration and may all be deleted.

A useful property of $\vdash$ may be stated

(3.22) If "$P \vdash Q$" is true and "$Q \vdash R$" is true, then "$P \vdash R$" is true.

To see this, suppose

$$S_1, S_2, \ldots, S_k$$

is a demonstration of "$P \vdash Q$" and that

$$S_1^*, S_2^*, \ldots, S_r^* \tag{3.23}$$

is a demonstration of "$Q \vdash R$". Then it turns out that,

$$S_1, S_2, \ldots, S_k, S_1^*, S_2^*, \ldots, S_r^* \tag{3.24}$$

is a demonstration of "$P \vdash R$". To verify this, we have to check the sequence against the definition of a demonstration. Clearly, "S_r^*" is the same as "R" and the "S_i" for $i = 1, \ldots, k$ satisfy the definition. There is no difficulty about any "S_j^*" which comes under one of the cases D_2, D_3, or D_4 in (3.23), because it will come under the same case in (3.24). But if "S_j^*" comes under D_1 in the sequence (3.23), then "S_j^*" is the same as "Q", and this "S_j^*" cannot come under the case D_1 in the sequence (3.24). However, we know that "S_k" is the same statement as "Q", so that in (3.24), "S_j^*" is just a repetition of "S_k" and thus comes under case D_4.

The preceding argument can be extended to justify the following generalization of (3.22):

(3.25) If "$A_1, \ldots, A_n \vdash B_1$" is true and
"$B_1, B_2, \ldots, B_m \vdash P$" is true, then
"$A_1, \ldots, A_n, B_2, \ldots, B_m \vdash P$" is true.

The following illustrative demonstration is based on an axiom for natural numbers called the "Closure Axiom". In the demonstration, we have to use statements of the form "x is a natural number" so often that it is useful to have an abbreviation. For "x is a natural number" we shall write "$x \varepsilon N$". Here "N" stands for the class of natural numbers, and "ε" is read "is an element of" or "is a member of".

The Closure Axiom for natural numbers is

$$A_1: (x \varepsilon N)(y \varepsilon N) \rightarrow (x + y \varepsilon N)$$

We wish to exhibit a demonstration proving

$$(3.26) \qquad A_1, a \varepsilon N, b \varepsilon N, c \varepsilon N \vdash (a + b) + c \varepsilon N$$

Demonstration	*Analysis*
$S_1: a \varepsilon N$	D_1
$S_2: b \varepsilon N$	D_1
$S_3: c \varepsilon N$	D_1
$S_4: (a \varepsilon N)(b \varepsilon N)$	D_3, conj inf
$S_5: (a \varepsilon N)(b \varepsilon N) \rightarrow (a + b \varepsilon N)$	D_1, closure axiom A_1
$S_6: a + b \varepsilon N$	D_3, modus ponens [S_4,S_5]
$S_7: (a + b \varepsilon N)(c \varepsilon N)$	D_3, conj inf [S_6,S_3]
$S_8: (a + b \varepsilon N)(c \varepsilon N) \rightarrow ((a + b) + c \varepsilon N)$	D_1, closure axiom A_1
$S_9: (a + b) + c \varepsilon N$	D_3, mod pon [S_7,S_8]
	Finally, S_9 is the statement to be deduced.

The column headed *Analysis* is not a part of the demonstration, but should be regarded as indicating how the corresponding statements of the demonstration satisfy the conditions for a demonstration.

Compare this nine-step demonstration of the simple statement (3.26) with the usual kind of proof, which might be written:

> Since a and b are natural numbers, so is $a + b$ a natural number by the closure axiom. Then since c is also a natural number, so is $(a + b) + c$ a natural number by the closure axiom.

While the foregoing argument is practical and convincing, it is far from being a demonstration. Acceptance of it as a proof is based on the belief that it indicates the existence of a demonstration. Indeed, this is the situation with respect to virtually all proofs in mathematics. If the simple statement (3.26) requires a nine-step demonstration, imagine the length of a demonstration of the binomial theorem, or of almost any theorem of plane geometry! It is not surprising that most of the arguments offered as proofs in mathematics are not demonstrations, but should be regarded as abbreviated demonstrations, or as a set of clues to a demonstration. The degree of abbreviation, or the number of the clues, depends upon the audience for whom the proof is intended. Proofs in current mathematical journals are often so brief that they are meaningful only to specialists. But, however brief the proof, it is understood that the author stands ready to fill in missing steps if the validity of the proof is questioned.

It is the difficult and important task of a geometry teacher to teach his students what a proof is, for it is in a geometry class that a student gets his first chance to learn what it means to prove something formally. It is not easy to define what should be meant by a proof in terms suitable for a beginning geometry student. The definition of a demonstration given in this section has the advantage of precision and simplicity. There is a clear distinction between the demonstration and the analysis of it. The demonstration consists of a string of statements each satisfying one of the conditions D_1, D_2, D_3, D_4 of the definition. The analysis simply indicates which condition each statement of the demonstration satisfies.

The precision and simplicity of the definition are achieved at the cost of great length in the demonstrations. A more complex definition might well lead to shorter demonstrations. In any event, the definition of this section seems too sophisticated for high school students, and complete demonstrations are too long and "hairsplitting" to present to them. Finally, it is a rare textbook that offers a list of axioms from which a demonstration, as defined here, can be constructed.

How then does a geometry student learn to construct acceptable proofs? He is taught to write out a proof as a series of statements supported by an analysis. He is taught that each statement must be supported by a "reason". As a guide to what statements are acceptable he may be instructed that each must be an instance of an axiom, a definition, a previously proved theorem, the hypothesis, etc. To describe acceptable "reasons" is not so easy. The "reasons" cannot always be simply descriptions of the statements to which they correspond. Often a "reason" is in fact a statement necessary to the inference of its corresponding statement. This is the case when "S.S.S.", or some other congruence theorem, is offered as a reason. Thus, the set of "reasons" is not pure analysis, but is, to some extent, part of the proof itself.

Since a precise definition of a proof is not generally available to him, the student must do a good deal of his learning by imitation. Gradually, through a process of observing examples of his teacher, or of his textbook, and by experimenting and adjusting to the authority of the teacher and the textbook, the student acquires an acceptable language of reasons, learns what to leave tacit, and begins to put together abbreviated demonstrations.

The geometry teacher has the delicate job of achieving the right compromise between formal rigor and the limitations of maturity and motivation of his students. To make the best compromise, he must not only understand the capabilities of his students but must also know what formal rigor is.

3.9 TREES

It will be useful in discussing methods of abbreviating a demonstration to exhibit the demonstration in the form of a *tree*. The *tree* form of the demonstration of (3.26) follows. [$S_1, S_2, \ldots, S_9$ are the statements in the sequence form of the demonstration of (3.26).]

(3.27)
$$\begin{array}{ll} 1. & \dfrac{S_1 \quad S_2}{S_4 \quad S_5} \\ 2. & \dfrac{}{S_6 \quad S_3} \\ 3. & \dfrac{}{S_7 \quad S_8} \\ 4. & \dfrac{}{S_9} \end{array}$$

1. Conj inf
2. Mod pon
3. Conj inf
4. Mod pon

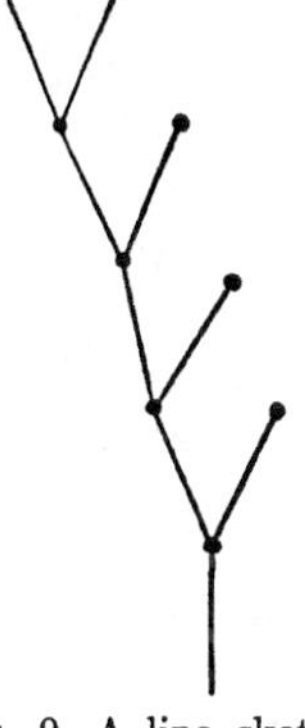

Fig. 9. A line sketch consisting of nine short line segments and four branch points.

In the tree (3.27) every horizontal bar indicates an inference whose designation is given in the right-hand column. Every statement occurring under a horizontal bar is inferred from the statements immediately above that bar. Statements that do not have a horizontal bar directly over them are of types D_1, D_2, D_4. The form of the tree (3.27) can be indicated by a simple line sketch, as in Fig. 9. In the sketch of Fig. 9, the nine short line segments represent the nine statements of the tree (3.27). Each of the four *branch points* represents an inference, and the segments leading downward from the branch points represent the statements "S_4, S_6, S_7, S_9", which are all of type D_3. The remaining segments, or *branches*, represent the statements "S_1, S_2, S_5, S_3, S_8", which are of types D_1, D_2, D_4. Of course, the bottom segment, or *trunk*, represents "S_9", the statement to be proved.

The tree form of a demonstration is not as compact as the sequence form, and may appear as quite an impressive structure if the theorem it demonstrates is at all complex. However, the tree form brings out clearly the role of inference in a demonstration, and is perhaps for this reason easier to understand than the sequence form.

3.10 ABBREVIATED DEMONSTRATIONS AS PROOFS

Recall that the statement (3.21)

(3.28) $$A_1, A_2, \ldots, A_n \models P$$

asserts the existence of a demonstration as defined in Sec. 3.8. To prove that (3.28) is true, it is enough to exhibit a suitable demonstration. We shall also consider as proof that "P" follows from "$A_1, \ldots, A_n$", any sound argument that shows the *existence* of a demonstration of (3.28), even though the demonstration is not actually exhibited. Such proofs arise when demonstrations are abbreviated by the use of previously proved statements.

In the definition of Sec. 3.8, no provision is made for using a previously proved statement as a step in the demonstration. This provision was avoided for the sake of a simple definition. But, if only demonstrations are allowed as proofs in plane geometry, for instance, then one is forced all the way back to the axioms in every proof. This would not be such a hardship in the case of the earlier theorems, but would require inordinately long proofs for the later theorems of geometry.

Assume that

$$A_1, A_2, \ldots, A_n \vdash B \tag{3.29}$$

had already been demonstrated, and it is desired to prove

$$A_1, A_2, \ldots, A_n \vdash P. \tag{3.30}$$

Suppose now that there exists a sequence of statements,

$$S_1, S_2, \ldots, S_k, \tag{3.31}$$

where "S_k" is the same as "P", and each "S_i" is one of "$A_1, A_2, \ldots, A_n$", or is in the form of a valid statement formula, or follows from earlier statements in the sequence by means of an established inference rule, or is the same as some earlier statement in the sequence or is the statement "B". If "B" occurs one or more times in the sequence (3.31), then (3.31) need not be a demonstration of (3.30), but it is clearly a demonstration of

$$B, A_1, A_2, \ldots, A_n \vdash P, \tag{3.32}$$

since then the occurrence of "B" in the sequence comes under the condition of being one of the statements "$B, A_1, A_2, \ldots, A_n$". From (3.29) and (3.32) it follows that there exists a demonstration of

$$A_1, A_2, \ldots, A_n, A_1, A_2, \ldots, A_n \vdash P. \tag{3.33}$$

To see this, in (3.25) replace, "B_1" by "B", "B_2" by "A_1", "B_3" by "A_2", . . . , "B_m" by "A_n".

It is easy to see that any demonstration of (3.33) is also a demonstration of (3.30). Thus, while the sequence (3.31) is not a demonstration of (3.30), it is a proof of (3.30) because it ensures that a demonstration of (3.30) exists.

We illustrate with a proof of a theorem of natural numbers, which depends on (3.26), proved in Sec. 3.8. Suppose we take as axioms for natural numbers, the closure, associative, and commutative axioms for addition, and the transitive axiom for equality:

A_1(clos.): $(x \in N)(y \in N) \rightarrow (x + y \in N)$
A_2(assoc.): $(x \in N)(y \in N)(z \in N) \rightarrow [(x + y) + z = x + (y + z)]$
A_3(comm.): $(x \in N)(y \in N) \rightarrow (x + y = y + x)$
A_4(trans.): $(x \in N)(y \in N)(z \in N)(x = y)(y = z) \rightarrow (x = z)$

Let us prove

(3.34) $a \in N,\ b \in N,\ c \in N,\ A_1,\ A_2,\ A_3,\ A_4 \vdash (a + b) + c = (b + c) + a$

Proof	*Analysis*
S_1: $a \in N$	Hyp
S_2: $b \in N$	Hyp
S_3: $c \in N$	Hyp
S_4: $(a \in N)(b \in N)(c \in N)$	Conj inf $[S_1,S_2,S_3]$
S_5: $(a \in N)(b \in N)(c \in N) \rightarrow$ $[(a + b) + c = a + (b + c)]$	Hyp, $[A_2]$
S_6: $(a + b) + c = a + (b + c)$	Mod pon $[S_4,S_5]$
S_7: $(b \in N)(c \in N)$	Conj inf $[S_2,S_3]$
S_8: $(b \in N)(c \in N) \rightarrow (b + c \in N)$	Hyp, $[A_1]$
S_9: $b + c \in N$	Mod pon $[S_7,S_8]$
S_{10}: $(a \in N)(b + c \in N)$	Conj inf $[S_1,S_9]$
S_{11}: $(a \in N)(b + c \in N) \rightarrow$ $[a + (b + c) = (b + c) + a]$	Hyp, $[A_3]$
S_{12}: $a + (b + c) = (b + c) + a$	Mod pon $[S_{10},S_{11}]$
S_{13}: $(a + b) + c \in N$	Instance of (3.26)
S_{14}: $(a \in N)(b + c \in N) \rightarrow [a + (b + c) \in N]$	Hyp, $[A_1]$
S_{15}: $a + (b + c) \in N$	Mod pon $[S_{10},S_{14}]$
S_{16}: $(b + c) + a \in N$	Instance of (3.26)
S_{17}: $[(a + b) + c \in N][a + (b + c) \in N]$ $[(b + c) + a \in N][(a + b) + c = a + (b + c)]$ $[a + (b + c) = (b + c) + a]$	Conj inf $[S_{13},S_{15},$ $S_{16},S_6,S_{12}]$
S_{18}: $S_{17} \rightarrow [(a + b) + c = (b + c) + a]$	Hyp, $[A_4]$
S_{19}: $(a + b) + c = (b + c) + a$	Mod pon $[S_{17},S_{18}]$

In spite of its length, the proof is not a demonstration because "S_{13}" and "S_{16}" do not satisfy any of the conditions D_1 (hypothesis), D_2 (form of a valid statement formula), D_3 (deduction from an established inference rule), or D_4 (a repetition of an earlier statement in the sequence) of the definition of a demonstration. But, if the first eight steps of the demonstration of (3.26) are inserted in the proof after "S_{12}", and the same steps (with an obvious change of notation) are inserted after "S_{15}", the resulting sequence of statements will be a demonstration. The demonstration will have sixteen more steps, and many repetitions.

While not a demonstration, the proof has a tree form:

$$\dfrac{\dfrac{\textcircled{S_{13}} \qquad \dfrac{\dfrac{S_1 \qquad \dfrac{\dfrac{S_2 \quad S_3}{S_7} \qquad S_8}{S_9}}{S_{10}} \qquad S_{14}}{S_{15}} \qquad \textcircled{S_{16}} \qquad \dfrac{\dfrac{S_2 \quad S_3}{S_4} \qquad S_5}{S_6} \qquad \dfrac{S_{11} \qquad \dfrac{S_1 \qquad \dfrac{\dfrac{S_2 \quad S_3}{S_7} \qquad S_8}{S_9}}{S_{10}}}{S_{12}}}{S_{17}} \qquad S_{18}}{S_{19}}$$

We have circled "S_{13}" and "S_{16}" since if the tree were a true demonstration, these statements would not be simple branches, but would have more branches above them, namely, tree forms of the demonstration of (3.26). The tree may be sketched as in Fig. 10.

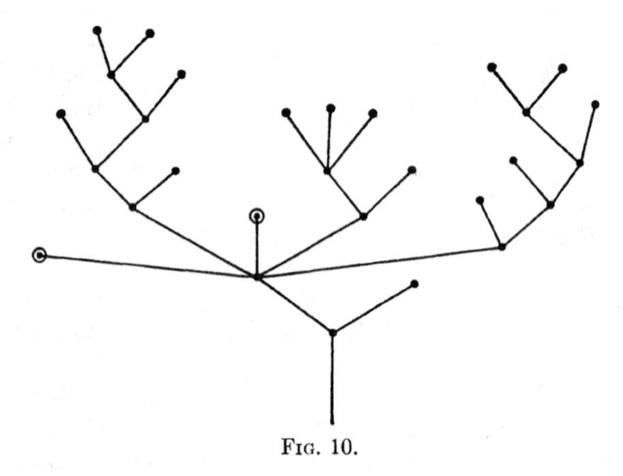

FIG. 10.

The sketch (Fig. 10) shows seventeen branches, two of which can be thought of as trunks of the system of branches in Fig. 9.

As a first abbreviation of the proof of (3.34), an obvious step is to omit the conclusions of conjunctive inferences. This type of abbreviation would eliminate "S_4, S_7, S_{10}, S_{17}". To leave out "S_4" where it first occurs in the proof is, in effect, to merge it with the modus ponens inference yielding "S_6". With these mergers modifying the tree it may be sketched in as in Fig. 11.

If now it is agreed to abbreviate by omitting the major premises of all modus ponens inferences, then "S_5, S_8, S_{11}, S_{14}, S_{18}" will be eliminated.

So much abbreviation might make the proof hard to follow, and so a clue to the tacit major premise will often be provided in the analysis column opposite the conclusion of the inference in question. With these abbreviations, the tree is further modified and its sketch appears in Fig. 12.

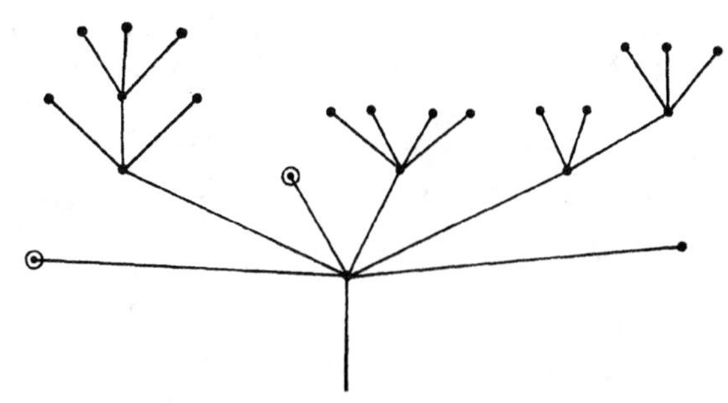

FIG. 11. Line sketch of Fig. 10, modified by eliminating certain line segments.

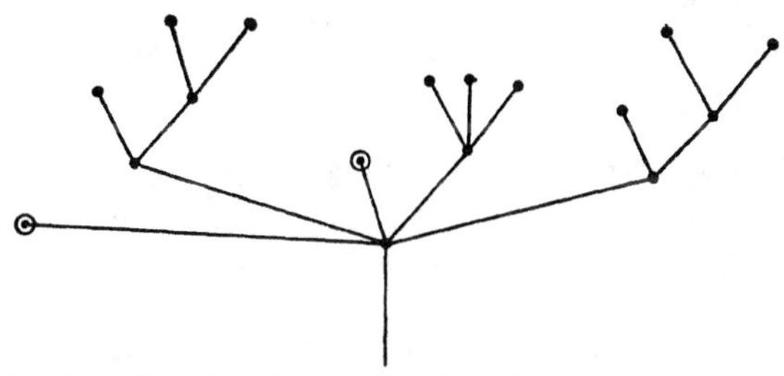

FIG. 12. Line sketch of Fig. 11 further modified by elimination of certain line segments.

The sequence form of the proof is reduced by these abbreviations to the following ten statements:

Proof	*Analysis*
S_1: $a \varepsilon N$	Hyp
S_2: $b \varepsilon N$	Hyp
S_3: $c \varepsilon N$	Hyp
S_6: $(a + b) + c = a + (b + c)$	Mod pon $[A_2, S_1, S_2, S_3]$
S_9: $b + c \varepsilon N$	Mod pon $[A_1, S_2, S_3]$
S_{12}: $a + (b + c) = (b + c) + a$	Mod pon $[A_3, S_1, S_9]$
S_{13}: $(a + b) + c \varepsilon N$	(3.26)
S_{15}: $a + (b + c) \varepsilon N$	Mod pon $[A_1, S_1, S_9]$
S_{16}: $(b + c) + a \varepsilon N$	(3.26)
S_{19}: $(a + b) + c = (b + c) + a$	Mod pon $[A_4, S_{13}, S_{15},$ $S_{16}, S_6, S_{12}]$

Another common type of abbreviation is the combination into one statement of several statements of the same kind. In the latest version of our abbreviated proof, we might choose to combine "S_1, S_2, S_3" and "S_9, S_{15}" and "S_{13}, S_{16}", respectively, into three statements, thus obtaining a proof having just the following six steps.

1. a, b, c are natural numbers	1. Hyp
2. $(a + b) + c = a + (b + c)$	2. Mod pon [1, assoc. ax.]
3. $b + c$ and $a + (b + c)$ are natural numbers	3. Mod pon [1, clos. ax.]
4. $a + (b + c) = (b + c) + a$	4. Mod pon [1,3, comm. ax.]
5. $(a + b) + c$ and $(b + c) + a$ are natural numbers	5. (3.26)
6. $(a + b) + c = (b + c) + a$	6. Mod pon [3, 5, 2, 4, trans. ax.]

This final form is about what would normally be offered as a careful proof of (3.34), although the analysis would not usually contain any reference to modus ponens.

3.11 DEDUCTION PRINCIPLE

It is well to observe in the preceding sample demonstrations exactly what was demonstrated. In (3.26), the existence of a demonstration shows that "$(a + b) + c \in N$" can be correctly deduced from the four statements

$$A_1, a \in N, b \in N, c \in N.$$

The demonstration did not demonstrate

$$A_1 \vdash (a \in N)(b \in N)(c \in N) \rightarrow [(a + b) + c \in N]. \tag{3.35}$$

Similarly, in dealing with (3.34),

$$a \in N, b \in N, c \in N, A_1, A_2, A_3, A_4 \vdash (a + b) + c = (b + c) + a,$$

we did *not* show that

$$A_1, A_2, A_3, A_4 \vdash (a \in N)(b \in N)(c \in N) \rightarrow [(a + b) + c = (b + c) + a]. \tag{3.36}$$

However, in both cases, it is (3.35) or (3.36) that we would really like to demonstrate. In general, we need to modify the foregoing notions of a demonstration in order to prove theorems. A theorem of geometry, for instance, is regarded as a statement that can be deduced from the axioms of geometry. Theorems of geometry are usually in the form of conditionals. Suppose "$P \rightarrow Q$" is the translation for a geometry theorem, and "$A_1, \ldots, A_n$" are translations for axioms. Then the usual problem is to prove that "$P \rightarrow Q$" can be deduced correctly from the axioms. What is wanted is a demonstration of

(3.37) $$A_1, \ldots, A_n \vdash P \rightarrow Q.$$

A typical procedure is to deduce "Q" from the axioms and "P", that is, to demonstrate

(3.38) $$A_1, \ldots, A_n, P \vdash Q.$$

Once (3.38) has been demonstrated, it is customary without further ado to assert that the theorem has been proved. The tacit proposition is "If (3.38), then (3.37)". The principle involved may be stated in general form as follows:

(3.39) **Deduction Principle:** If $A_1, A_2, \ldots, A_n, P \vdash Q$, then $A_1, A_2, \ldots, A_n \vdash P \rightarrow Q$.

While we shall take (3.39) as a logical principle with no attempt at proof, it should be noted that (3.39) is in fact a theorem of logic, and proofs of it can be found in current textbooks on mathematical logic.[1] The form (3.39) asserts that if there is a demonstration $S_1, S_2, \ldots, S_r$ of $A_1, \ldots, A_n, P \vdash Q$, then there exists also a demonstration $S_1^*, S_2^*, \ldots, S_s^*$ of $A_1, \ldots, A_n \vdash P \rightarrow Q$. The usual proof consists in showing how to construct the S^*'s given the S's.

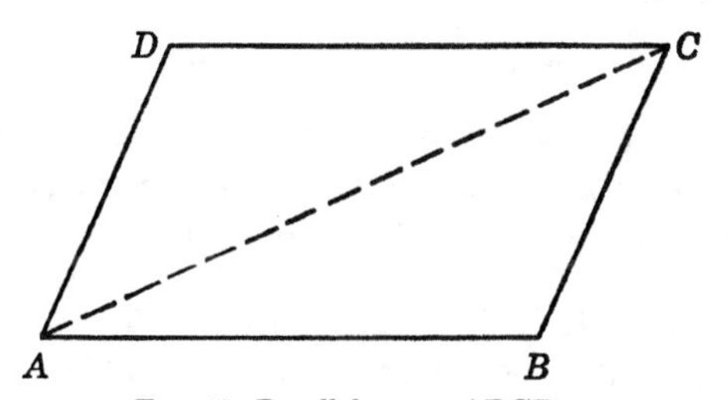

Fig. 13. Parallelogram ABCD.

As an example of application of the deduction principle consider the plane geometry theorem:

(3.40) If a figure is a quadrilateral, then, whenever its opposite sides are equal, it is a parallelogram.

Using the notation of Fig. 13, let "P", "Q", "R" be translations as follows:

[1] J. Barkley Rosser, *Logic for Mathematicians*, p. 75, New York: McGraw-Hill Book Company, Inc., 1953. Alonzo Church, *Introduction to Mathematical Logic, Part I*, p. 10, Princeton, N.J.: Princeton University Press, 1944. Stephen Cole Kleene, *Introduction to Metamathematics*, p. 90, Princeton, N.J.: D. Van Nostrand Company, Inc., 1952.

P: ABCD is a quadrilateral.
Q: AD = BC and AB = DC.
R: ABCD is a parallelogram.

Then we take the following instance of (3.40) as the theorem to be proved:

(3.41) $$P \rightarrow (Q \rightarrow R).$$

The usual proof begins with the assumption of "P" and "Q", and goes on to show that the diagonal AC divides the quadrilateral into two congruent triangles, and from this result deduces that the opposite sides are parallel, and hence "R". Such a proof establishes

(3.42) $$\text{Axioms}, P, Q \vdash R,$$

but it is (3.41) that we desire to deduce from the axioms. It is at this stage that the deduction principle may be used to obtain

(3.43) $$\text{Axioms}, P \vdash Q \rightarrow R.$$

Note that we have not actually produced a demonstration of (3.43), but assert on the basis of existence of a demonstration of (3.42) and the deduction principle that a demonstration of (3.43) *exists*. However, (3.43) is still not what we want. We have to use the deduction principle once more to obtain

(3.44) $$\text{Axioms} \vdash P \rightarrow (Q \rightarrow R).$$

Use of the deduction principle is tacit in virtually all proofs in plane geometry. Surely a high school student who succeeded in demonstrating (3.42) would be happy to write Q.E.D. and would not be conscious of the need to say anything more. A plane geometry student bent on proving a theorem in the form "$H \rightarrow C$" almost surely looks upon the conclusion "C" as the thing he is trying to prove true. While a mathematician is quite clear that it is "$H \rightarrow C$" that he is trying to prove, he normally uses the same procedure as the high school student, namely, he assumes "H" and gives a demonstration for

(3.45) $$\text{Axioms}, H \vdash C,$$

and for him also the deduction principle is tacit in the proof of

(3.46) $$\text{Axioms} \vdash H \rightarrow C.$$

It is only when one tries to give precision to the notion of a proof that the distinction between (3.45) and (3.46) becomes urgent. In many cases in mathematics, care must be exercised in the use of the deduction principle, and we shall have to reconsider it in a later section when we deal explicitly with proofs of quantified statements.

3.12 CONVENTIONAL PROOFS IN GEOMETRY

The conventional proof of a theorem of geometry has the following pattern:

1. The theorem is stated.
2. A figure is drawn and lettered.
3. The theorem is restated in terms of the notation of the figure, as "$P \rightarrow Q$" or often in the form: Given: "P"; To Prove: "Q".
4. An abbreviated demonstration is given of "Axioms, $P \vdash Q$".
5. By a tacit use of the deduction principle it is considered that "Axioms $\vdash P \rightarrow Q$" has been proved.
6. Although "P" and "Q" are statements about the particular lines and points of Step 3, it is assumed that Steps 2 to 5 are proof of the theorem, which is a statement about all such lines and points.

Let us investigate conventional proofs with respect to Steps 2 to 4. Among the abbreviations occurring in Step 4 are found the types mentioned in Sec. 3.10, namely,

a. Use of proved theorems
b. Omission of conclusions of conjunctive inference
c. Omission of the major premise in modus ponens inference
d. Combination of steps of the same type

However, additional abbreviations are commonly employed, and certain conventional words and phrases occur in the analysis column.

As we have seen (Sec. 3.8), a demonstration consists of a sequence of statements satisfying certain formal requirements. The analysis of a demonstration is not essential to it, and consists of comments about the form of the statements in the demonstration. In contrast, the analysis column of the usual abbreviated proof in geometry contains statements essential to the proof and is not just a sequence of comments on the form of the statements in the "statement" column. If then, such a proof is not to be hopelessly abbreviated, it must be accompanied by an analysis, or a column of "reasons". However, the analysis also serves a pedagogical purpose in that it helps a teacher to determine to what extent a student knows what he is doing. Furthermore, writing out definitions and theorems and axioms for the analysis helps a student learn these theorems and definitions and axioms. For these reasons, it is commonly the "reason" column that is emphasized in teaching students how to write out proofs. They are often given explicit instructions about the statement to which the "reason" corresponds. With the admonition "Every statement must have a reason" ringing in his ears, a student thinks of the "reason" as something that assures the *truth* of the corresponding statement. This is not at all the role of the analysis column of a demonstration as described in Sec. 3.8.

It is not really very hard to write out proofs that, even though abbreviated, are more in the spirit of demonstrations. It is only necessary to transfer some statements traditionally appearing in the reason column to the statement column. Let us examine a typical conventional proof and see how this can be done. We shall follow the pattern for a conventional proof in geometry given on page 91.

Theorem: In a parallelogram, the diagonals bisect each other.

Draw a figure (Fig. 14) and letter it. State the theorem in terms of the notation of the figure and in the "Hypothesis-Conclusion" form:

Hypothesis: ABCD is a parallelogram with the diagonals AC and BD meeting at M[1] (Fig. 14).

Conclusion: AM = MC and BM = MD.

We are now ready to embark on Step 4 of the pattern of proof, page 91.

Proof I

Statements	*Reasons*
1. ABCD is a parallelogram with opposite sides AB and DC.	1. Hypothesis.
2. AB ∥ DC.	2. Definition: Opposite sides of a parallelogram are parallel.
3. ∠CAB = ∠DCA.	3. Theorem: If two parallel lines are cut by a transversal, alternate interior angles are equal.
4. Similarly, ∠ABD = ∠BDC.	4. See Step 3.
5. AB = DC.	5. Theorem: Opposite sides of a parallelogram are equal.
6. ∴ △ABM ≅ △DCM.	6. Two triangles are congruent if two angles and the included side of one are equal, respectively, to two angles and the included side of the other.
7. Hence, AM = MC, BM = MD.	7. Definition: Corresponding parts of congruent triangles are equal.

Q.E.D.

[1] Of course, we are not really dealing with an instance of the theorem in question, because the "intersection of the diagonals" is part of the conclusion. However, it is a rare geometry book that supplies the necessary axioms to prove this, so we shall not attempt it here. In fact, we are proving "if the diagonals of a parallelogram intersect, then they bisect each other".

Note the further abbreviation that leaves tacit the proof that ABM and DCM are triangles.

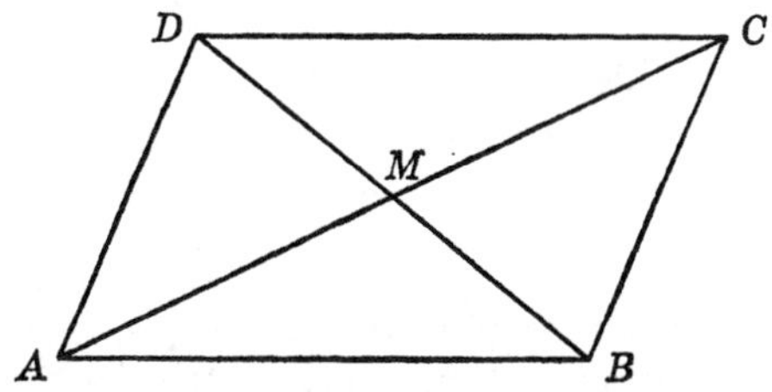

FIG. 14. Parallelogram ABCD with diagonals AC and BD intersecting in M.

Now let us rewrite the proof by shifting the statements and definitions over into the statement column. This involves no more writing than before, except that the use of modus ponens is indicated.

Proof II

Statements	*Analysis*
1. ABCD is a parallelogram with opposite sides AB and DC.	1. Hypothesis.
2. Quadrilateral ABCD is a parallelogram if and only if its opposite sides are parallel (AB ∥ DC, AD ∥ BC).	2. *Instance of a* definition.
3. AB ∥ DC.	3. *Mod pon* [1,2].
4. If transversal AC cuts the parallels DC and AB, then ∠CAB = ∠DCA.	4. *Instance of a* theorem.
5. ∠CAB = ∠DCA.	5. *Mod pon* [3,4].
6. ∠ABD = ∠BDC.	6. Similarly.
7. If AB and DC are opposite sides of parallelogram ABCD, then AB = DC.	7. *Instance of a* theorem.
8. AB = DC.	8. *Mod pon* [1,7].
9. Triangles ABM and DCM are congruent if ∠CAB = ∠DCA, AB = DC, and ∠ABD = ∠BDC.	9. *Instance of a* theorem.
10. △ABM ≅ △DCM.	10. *Mod pon* [5,6,8,9].
11. Triangles ABM and DCM are congruent if and only if their corresponding parts are equal.	11. *Instance of a* definition.
12. AM = MC, BM = MD.	12. *Mod pon* [10,11].

In comparing these proofs, notice that the extra words of Proof II are in italics. It is worth observing that there are few words in Proof II not appearing in Proof I, but Proof I makes no reference to modus ponens. If in Proof II, modus ponens is to be tacit, then each of Steps 3,5,8,10, and 12 would be merged with its preceding step, thus yielding a seven-step proof as in Proof I. For instance, Steps 3 and 2 might be merged as follows:

2. Quadrilateral ABCD is a parallelogram if and only if its opposite sides are parallel. $\therefore$ AB $\parallel$ DC	2. Definition

If such mergers are to be considered proper, it may be helpful to make an agreement that " $\therefore$ " always introduces the conclusion of a modus ponens.

Proof II as it stands, without these mergers, is not really any longer than Proof I. In particular, *every* statement in the statement column of Proof II occurs somewhere in Proof I. In Proof II, the analysis is truly descriptive and not at all necessary to the proof. Also, the analysis is not as necessary pedagogically as is the set of reasons in Proof I. If a student wrote out the set of statements 1 to 12 of Proof II without any accompanying analysis, would you not be inclined to believe that he knew what he was doing?

In the analysis of Proof II, consider Step 7: there we say "instance of a theorem" rather than just "theorem", because the theorem in question is

The opposite sides of a parallelogram are equal,

which is a general statement about all parallelograms, whereas the statement in Step 7 is a statement about ABCD which, though unknown, is a fixed particular parallelogram. Of course, the statement in Step 7 is true, because the theorem is true for all parallelograms.

Proof II is not only easier to follow than Proof I, but is much easier to describe. Every statement in the statement column is either the conclusion of a modus ponens from earlier statements, or a statement considered true because it is part of the hypothesis, or an instance of a theorem, an axiom, or a definition. The analysis simply notes that the statements have these properties. As an aid in following the proof, the analysis of a modus ponens conclusion indicates the major and minor premises by numbers. No such simple description is possible for Proof I.

As the geometry course progresses and it is felt that abbreviations are desirable, Proof II can be abbreviated as easily as Proof I. If, for instance, it is desired to use "A.S.A." to indicate the congruence theorem in Proof I, Step 6, it can as well be used in Proof II, Step 9.

3.13 ANALYSIS OF A CONVENTIONAL PROOF

Now, let us consider the relationship between a theorem and its corresponding figure. In a sense, we can regard the streaks and dots of chalk we draw on a blackboard as real interpretations or models of the abstract lines and points of geometry. This is probably our attitude when carrying out straightedge and compass constructions. However, when we draw a figure in connection with a geometric proof, it is probably safer to regard it as a symbol, or a collection of symbols, naming the lines and points with which the theorem is concerned. In proving the theorem on page 92, it is necessary to talk about the parallelogram and its diagonals. In order to talk about the parallelogram, we need a name for it, such as "the parallelogram". In fact, we used "ABCD" as a name for the parallelogram. This is a fairly good name, because we also used "A", "B", "C", and "D" as names for the vertices of the parallelogram, and this relationship between the names of the vertices and of the parallelogram helps us to think about them more clearly. Figure 14, including the letters, contains a name for the parallelogram, as well as names for vertices, diagonals, sides, and so on. Let us imagine that the streaks of ink naming diagonals are removed from the figure, along with the letter "M", then, what is left is a symbol or name for the parallelogram. Since this figure and "ABCD" are both names of the same parallelogram, they could be used interchangeably in any statement about the parallelogram. The figure is, perhaps, a better name for the parallelogram than is "ABCD" because it carries more information for us. From the figure we would expect to name one side of the parallelogram "AB", for instance, while we would not have that expectation just from the name "ABCD".

It is not normally very convenient to use pictorial names in written statements. One does not write

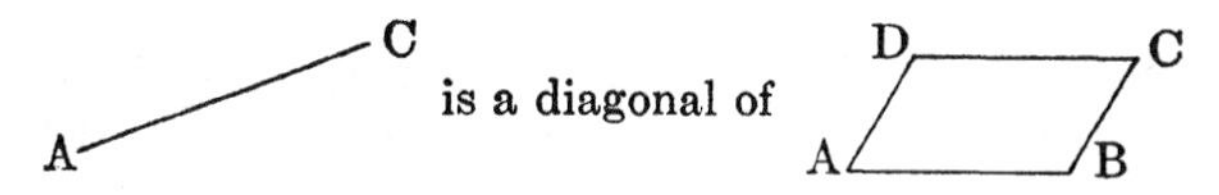

but rather one writes

AC is a diagonal of ABCD.

The figure is not used in statements, but is nevertheless useful because it can be thought of as a collection of names carrying more accessible information than the literal names used in statements. For example, in the statement

$$\angle DAB = \angle BCD,$$

the angles are named by the literal symbols "$\angle$DAB" and "$\angle$BCD". It is surely helpful to look at the corresponding names for these angles to be found in the figure. The figure suggests relations between these angles and other parts of the parallelogram. In addition, the figure aids in comprehending the statement of equality of angles in relation to other statements that might be made about the parallelogram.

This use of a figure is a two-edged affair. Relations existing between the parts of the figure do not necessarily correspond to relations between the parts of the parallelogram. Anything suggested about the parallelogram by its figure must be verified. But the pictorial names, or figures, used in geometry are very clever names and their suggestions are often so powerful that one sometimes accepts them without proof. In the proof of the theorem, page 92, the figure suggests so strongly that AC and DB intersect, that not one student in a thousand feels any need for proving it. Literal symbols are not so seductive. One is not, for instance, inclined to assert that the diagonal AC contains two letters just because its name "AC" contains two letters. It is perhaps for these reasons that both literal and pictorial names are used. The pictorial symbols are useful because they are highly suggestive but are for this reason dangerous. To verify the suggested relations, one writes out proofs using literal symbols to avoid the danger.

A geometry teacher has a delicate job here. On the one hand, he has to train a student to make the utmost use of figures, but on the other hand he has to be sure that the student does not identify the figure with the geometric abstraction of which it is a name. The conventional language of geometry tends to encourage identification of a pictorial name with the thing it names. For instance, the same word "line" may be used to name the undefined geometric element and the chalk streak or ink deposit of a figure. The direction

Draw the line AC

may suggest to the student that he is supposed to create a line joining A and C, which points presumably are not at the moment joined by any line. This is, of course, absurd. One can, however, draw a *picture* of the line AC.

It is common practice in some textbooks to state as an axiom:

One and only one straight line can
be drawn through any two points.

This is not at all intended to be a statement about figures or pictures, but is supposed to be an assumption about the abstractions *line* and *point.* The language is misleading, but common. It would be far better to state the axiom:

There exists one and only one straight line
through any two points.

The language of construction problems is especially loaded with expressions that incline one to think about the figure rather than the abstraction for which the figure stands. In particular, it is common practice to have geometry students describe a construction just after stating a theorem in the Given–To Prove form and before embarking on the proof. Then, in the proof, "construction" may be offered as a reason for a step. If the construction is really a description of the figure, then "construction" can never be offered as a reason for a statement in the proof of the theorem, for such a statement is never an assertion about the figure. Statements used in the proof of the theorem must be statements concerning the abstract configuration of points and lines, like triangle or parallelogram, mentioned in the theorem and not statements about the figure. The use of the word "construction" might be justified if the existence of the lines and points constructed is proved, but such proof belongs with the rest of the proof and seems out of place if given as a preliminary. Consider a conventional proof of a theorem involving an isosceles triangle, where it is understood that all the usual congruence theorems about triangles have been assumed as axioms.

Theorem: The base angles of an isosceles triangle are equal.

First we draw a figure (Fig. 15) and letter it.

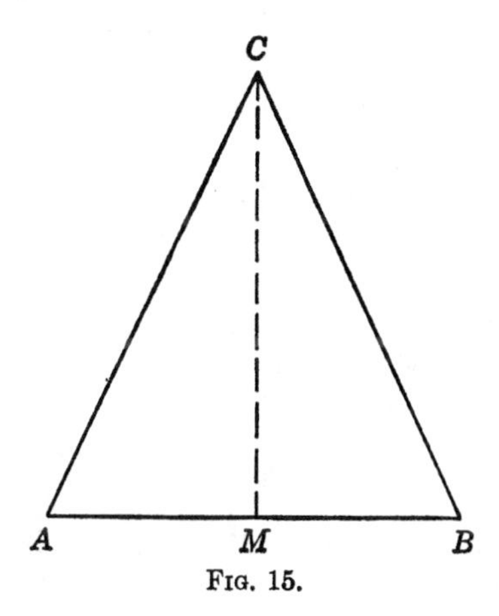

FIG. 15.

Then we proceed as follows:

Hypothesis: $\triangle$ABC is isosceles with AC = BC (Fig. 15).
Conclusion: $\angle$CAB = $\angle$CBA.
Construction: Draw CM joining vertex C with the mid-point M of AB.

Proof

Statement	*Reason*
1. △ABC with AC = BC.	1. Hypothesis.
2. AM = BM.	2. Construction.
3. CM = CM.	3. Identity.
4. ∴ △AMC ≅ △BMC.	4. Axiom: Two triangles are congruent if the three sides of one are equal respectively to the three sides of the other.
5. Hence ∠CAB = ∠CBA.	5. Corresponding parts of congruent triangles are equal.

The proof is incomplete because the construction has not been proved.[1] If the construction is to be justified, its proof might as well occur in the main body of the proof, perhaps as follows:

Hypothesis: △ABC is isosceles with AC = BC (Fig. 15).
Conclusion: ∠CAB = ∠CBA.

Proof

Statement	*Analysis*
1. △ABC with AC = BC.	1. Hypothesis.
2. There is a point between A and B, call it "M", such that AM = BM.	2. Instance of an axiom.
3. There exists a unique line through C and M.	3. Instance of an axiom.
4. CM = CM.	4. Identity.
5. In triangles AMC and BMC, if AC = BC, CM = CM, and AM = BM, then △AMC ≅ △BMC.	5. Instance of an axiom.
6. ∴ △AMC ≅ △BMC.	6. Mod pon [1,2,4,5].
7. Two triangles are congruent, namely, AMC and BMC, if and only if their corresponding parts are equal.	7. Instance of a definition.
8. ∴ ∠CAB = ∠CBA.	8. Mod pon [7,6].

The use of the word "identity" in both proofs is worth a comment. In order to be of use in proving the congruence of triangles AMC and BMC, one must interpret "CM = CM" as asserting "The length of the

[1] We might also note that, though quite obvious, it was not established that AMC and BMC are triangles. To show them triangles, we might argue: A, M, and B are collinear by definition. If A, M, and C are collinear, then A, B, and C are collinear since there is a unique line through A and M. But A, B, and C are not collinear by the definition of a triangle. Therefore A, M, and C are not collinear by contrapositive inference, and AMC is a triangle by definition.

side CM of triangle AMC is equal to the length of side CM of triangle BMC". Of course, these sides are the same line segment CM, and the equality is an instance of an axiom. All of this is abbreviated in the statement "CM = CM" and the reason "identity".

EXERCISES

Look up proofs of the following statements or make your own proofs. Try to write each proof in the spirit of Proof II, Sec. 3.12, with an analysis column that is "pure" analysis.

1. Prove: The sum of an even number and an odd number is odd.

2. Prove: A diagonal of a parallelogram divides it into two congruent triangles.

3. Prove: The sum of two even numbers is even.

4. Prove: $\sin^2 \theta + \cos^2 \theta = 1$.

5. Prove the law of sines.

6. Prove: If n^2 is even and n is an integer, then n is even.

7. Prove: In a circle, an angle formed by a tangent and a chord drawn to the point of contact is measured by one-half of the intercepted arc.

8. Prove: $(ab = 0) \rightarrow (a = 0) \vee (b = 0)$, where a and b are real numbers.

9. Prove: A point is on the bisector of an angle if and only if it is equidistant from the sides of the angle.

10. Prove: The bisectors of the interior angles of a triangle are concurrent.

IV: ABSTRACT MATHEMATICAL SYSTEMS

4.1 MATHEMATICAL MODELS

It is a revealing experiment to pose the question: "What is a point?" to a mathematics class. Typical responses are: "A point is a position in space", "A point has no length or width", "A point is the intersection of two lines", "A point has location but no dimensions". It is easy enough to point out how each of these attempts fails as a definition. If now the questioner persists in posing the question, the responses dwindle, and the class may well fall into an angry frustrated silence. What is of interest here is that only rarely will any student take issue with the legitimacy of the question. It seems quite proper to the student that after studying plane geometry for a year he ought to be able to define a *point* in the same sense that he is able to define a *triangle* or *circle.*

Of course, the question is not legitimate since the word "point" in geometry is not defined in the usual sense at all. The words "point", "line", "distance", "on", and some others, are primitive notions or undefined words of the system. They are taken as undefined in order to avoid circularity of definitions.

It seems intuitively clear that any attempt to define every geometric word must lead to circularity of some sort. Consider, for instance, an immense dictionary containing all the one-half million or more words in the English language. Each word in this dictionary is defined in terms of other English words, also of necessity in the dictionary. It seems inevitable if one starts to track down the definition of "point" by looking up its definition, then looking up all the words in its definition and so on, that one sooner or later comes upon a word whose definition uses the word "point", and a circularity has occurred.

On the other hand, given a set of undefined words of geometry designated by such symbols as "point", "line", "distance", "on", "between", it will be possible to define further words in terms of these undefined words. It is possible to compose statements using these words, and it may be possible to order these statements into a system in which it can be shown that some of the statements follow logically from a small subset regarded as axioms. Such a program presupposes a basic language and requires a system of logic that describes how a statement *follows* from other statements.

The familiar axiom of Euclidean geometry,

Through two distinct *points* there is one and only one *line*,

is a mixture of the underlying language (in this case English) and primitive notions or undefined words of geometry, which have been italicized. In viewing Euclidean geometry as a formal system, it must be emphasized that the primitive notions must be regarded as meaningless. A surveyor may regard the cross hairs in the telescope of his transit as a *point* and may regard his line of sight as a *line*, or regard a string held taut by a plumb bob as a *line*. To a carpenter, a *line* may be in practice the edge of a plank or a streak of chalk deposited on a board by snapping a chalk-laden string. *Point* and *line* to a draftsman may be a dot of ink and a streak of ink made by a pen on drafting paper. All of these so-called lines and points have many properties not attributed to the lines and points of geometry and must be regarded as pictures or interpretations of geometric lines and points.

To put it another way, the carpenter, the surveyor, and the draftsman each has his own interpretation of the axioms of Euclidean geometry. Each finds Euclidean geometry to be a satisfactory body of statements about the behavior of points, lines, and so on, as he interprets these words. But Euclidean geometry as a formal system has to be considered as uninterpreted.

Note that even though uninterpreted, the axioms of geometry tell us all we need to know about the properties of the undefined words of geometry. While the axioms do not tell us what the undefined words mean, they do define the properties of the undefined words in the sense that any statement about them is proved or disproved on the basis of the axioms.

The notion of an abstract mathematical system can be summarized as follows. The basic elements are:

1. An underlying language
2. A deductive logic system
3. A vocabulary of undefined words
4. A set of axioms—statements about the undefined words
5. Theorems—statements about the undefined words that can be demonstrated, as in Secs. 3.8 to 3.11

Before going on to an example of an abstract mathematical system, a miniature geometry, a comment about definitions is, perhaps, in order. In an uninterpreted abstract system, a definition is a statement of equivalence that is used to introduce new, and generally shorter, names for more complex statements. In theory, an abstract system can be built without definitions but with a great loss of economy in use of words. In plane geometry, for instance, instead of introducing the name "triangle", the description "three points not on the same line and the three line segments determined by them" might be used.

Definitions in mathematics often appear in the form of conditional statements, but nevertheless they are always to be taken as statements of equivalence. Although the statement,

A triangle is *isosceles* if it has two equal sides,

does not say so, it is to be understood that "if a triangle is isosceles, it has two equal sides". The definition is correctly stated:

A triangle is isosceles if and only if it has two equal sides.

The form of definition in which the "only if" remains tacit is quite common in higher mathematics and causes no trouble there. In elementary mathematics, where some students have a tendency to treat all conditionals as statements of equivalence, it might be wise to state definitions in the if-and-only-if form.

4.2 A MINIATURE GEOMETRY

To illustrate the notion of an abstract mathematical system, we shall describe one having just three undefined words and six axioms. The resulting structure is limited enough to view as a whole and yet has enough complexity to serve as a fair model of all abstract mathematical systems.

Before we begin a description of the system, a note on notation is in order. Axioms will be designated by a capital "A" and a subscript, definitions by a capital "D" and a subscript, and theorems by a capital "T" and a subscript. Lower-case letters will be used to refer to lines and upper-case letters to points.

Undefined words: "point", "line", "on".

Axioms and definitions:

A_0: Point P is on line q if and only if line q is on point P.
A_1: There exists at least one line.
A_2: There are exactly three distinct points on every line.

D_1: A number of points are said to be *joined* by a line s if and only if all the points are on s.

D_1^*: A number of lines are said to be *joined* by a point S if and only if all the lines are on S.

D_2: A point P is called a *pole* of a line p if and only if P is not joined to any point on p by a line.

D_2^*: A line p is called a *polar* of a point P if and only if p is not joined to any line on P by a point.

NOTE: The pole-polar relation will be frequently indicated by using the same letters, upper and lower case.

A_3: Every line has exactly one pole.
A_4: Every point has exactly one polar.
A_5: If a point P is not on line x but is joined to a point on x by a line m, then P is joined to exactly one other point on x by a line n distinct from m.

While we have named the undefined words "point", "line", "on", recall that the undefined words are meaningless and might as well be called by any other names, say "goo", "oog", "over". With these names, Axiom 2 would read:

There are exactly three distinct goos over every oog.

It will be easier to follow the development of theorems in the system if we continue to use the more familiar words "point", "line", "on", but we emphasize that the words are meaningless and must not be taken as names of real objects or relations. Furthermore, it will not do to confuse these points and lines with the points and lines of Euclidean geometry. From Axiom A_2, for example, it is clear that our lines, at least, are not Euclidean lines. The properties of our points and lines are given by the axioms, just as the properties of Euclidean points and lines are given by the Euclidean axioms.

The proofs that follow will not be complete demonstrations as described in Sec. 3.8, since even in this miniature system complete demonstrations are prohibitively lengthy. The proofs will be abbreviated but will illustrate many of the formal notions so far developed. The proofs will not be displayed in column form, but the source of each statement will be indicated by naming the axiom, theorem, or definition that applies. The name will appear inside square brackets following the statement.

Preceding the statement and proof of the first theorem of our system are two lemmas. A lemma is a theorem; however, it is a theorem whose interest lies not so much in its result but in its application in the proofs of other theorems whose results are of more moment, or of greater interest.

Lemma 1: Point P is a pole of line p if and only if line p is a polar of point P.

Proof: By using the notational convention regarding pole-polar relations, the lemma may be written

$$\text{P is a pole of p} \leftrightarrow \text{p is a polar of P.}$$

If the left member is translated as "A" and the right member as "B", the lemma can be written "$A \leftrightarrow B$".

Proof of "$A \rightarrow B$": Use the contrapositive method of proof, that is, prove "$\sim B \rightarrow \sim A$". Suppose p is not the polar of P (this is supposing

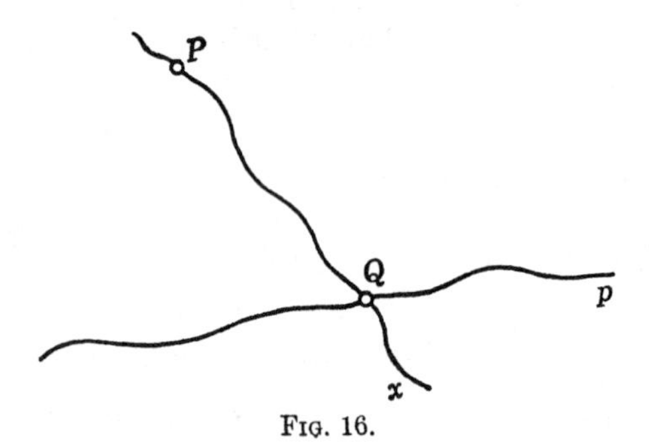

FIG. 16.

"$\sim B$"). Then p is joined to some line x on P by a point, call it "Q" (Fig. 16) [D_2*]. But then P is joined to Q on p by the line x [A_0,D_1]. Therefore, P is not a pole of p [D_2].

What has been established here is "$A_0, \ldots, A_5, \sim B \vdash \sim A$"; hence, by the deduction principle, we have "$A_0, \ldots, A_5 \vdash \sim B \rightarrow \sim A$". Then, by contrapositive equivalence (2.42) and substitution, Sec. 2.11, we are assured

$$A_0, \ldots, A_5 \vdash A \rightarrow B.$$

Proof of "$B \rightarrow A$": The assertion can be proved by showing "$\sim A \rightarrow \sim B$", and is left as an exercise.

Lemma 2: If point P is not joined to two points on p, then P is a pole of p.

Proof: The lemma is a statement in the form "$A \rightarrow B$". Again we employ a contrapositive form of proof, starting with "$\sim B$".

Suppose P is not a pole of p. Then P is joined to some point X on p by a line [D_2]. Now, P is on p, or P is not on p [an instance of the valid formula (2.36)].

Case I: P is joined to X on p by a line, and P is on p.

In addition to P and X on p, there is a third point on p [A_2], call it "Y". But P is joined to Y by p [D_1]; so we have that P is joined to two points on p, contradicting the hypothesis "A", that is, that P is not joined to two points on p.

Case II: P is joined to X on p by a line, and P is not on p.

With the assumption of Case II as minor premise, the axiom "A_5" as major premise, we have by modus ponens that P is joined to some point on p other than X. So again, P is joined to two points on p, contradicting the hypothesis "A".

It follows that "$\sim B$" leads to "$\sim A$", and we conclude as in the proof of Lemma 1.

It is interesting to spell out the logical form of this proof, as it is a variant of a proof by cases. We started with

1. "$\sim B$"

and, using the definition of a pole, got

2. "$\sim B \rightarrow C$"

So

3. "C" by modus ponens

Now,

4. "$S \vee \sim S$" is a valid formula,

so

5. "$C(S \vee \sim S)$" by conjunctive inference. (This was tacit in the proof.)

In Case I we proved

6. "$CS \rightarrow \sim A$".

In Case II we proved

7. "$C{\sim}S \rightarrow {\sim}A$",

so

8. "$(CS \vee C{\sim}S) \rightarrow {\sim}A$". Inference by cases (2.78).

But,

9. "$C(S \vee {\sim}S) \leftrightarrow (CS \vee C{\sim}S)$" by the distribution formula (2.48).

So, by substitution,

10. "$C(S \vee {\sim}S) \rightarrow {\sim}A$",

and so

11. "${\sim}A$" by modus ponens [5,10].

Steps 1 to 11 show that "${\sim}B$" leads to "${\sim}A$" and the proof concludes as in the proof of Lemma 1.

T_1: Two distinct points are not joined by more than one line.

Proof: The form of the proof is indirect. We start with "${\sim}T_1$" and show that this leads to a denial of "A_4", that is, to "${\sim}A_4$".

Suppose that there are two distinct points A and B, and A and B are joined by more than one line (Fig. 17). Then there are at least two distinct lines, m and n, joining A and B.

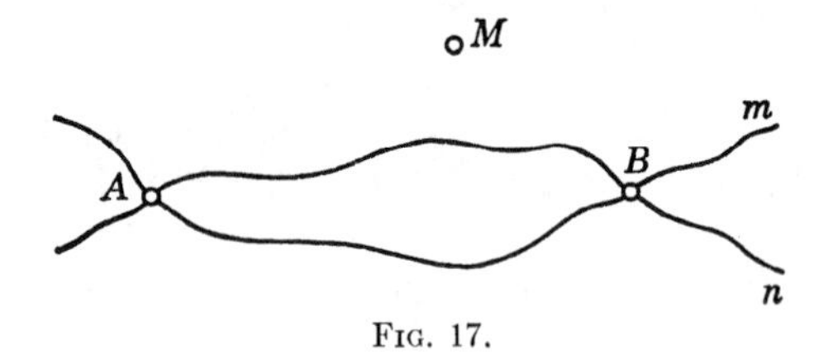

FIG. 17.

Now, line m has a unique pole M [A_3]. But, since M is not joined to A or B on n, M is also a pole of n [Lemma 2].

Then m and n are polars of M [Lemma 1], and since m and n are distinct, we have a denial of "A_4", that is, we have established "${\sim}A_4$", and the indirect proof is complete.

Theorem T_1 is also a true theorem about Euclidean points and lines. That is, the same statement follows from the axioms of Euclidean geometry. About Euclidean points and lines we can say, in addition to "T_1",

Every two points are joined by a line.

For the points and lines of our geometry this statement is false, for if point P is the pole of line p, then P is not joined to any of the three points on p; so there are pairs of points in our geometry not joined by any line.

With "T_1" we have established that if a pair of points are joined by a line, then the line is unique. Clearly there *are* pairs of points that are joined [A_1,A_2]. The next theorem throws light on the pole-polar relation.

T_2: If P is the pole of p, and Q is the pole of q, and P is on q, then Q is on p.

Proof: Using the notational convention for indicating the pole-polar relation, we may write the statement "T_2" in the shorter form

$$(\text{P on q}) \rightarrow (\text{Q on p}).$$

Since P is on q [hyp], there exist exactly two other points on q [A_2], call them "A" and "B" (Fig. 18).

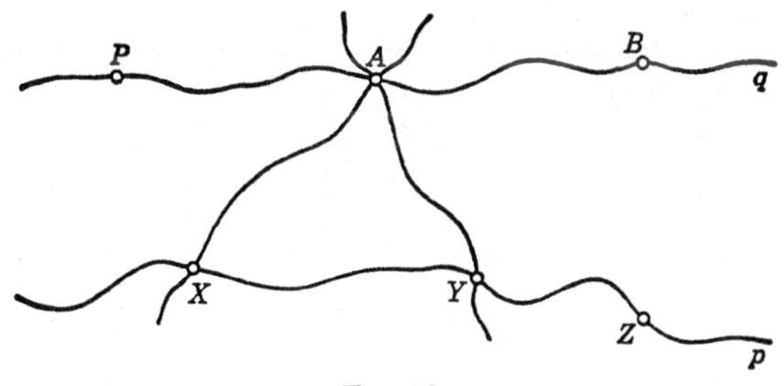

FIG. 18.

P is the pole of p [hyp]; therefore A is not the pole of p [A_3]. Hence, A is joined to some point on p [D_2]; call it "X". A is also joined to exactly one other point on p [A_5]; call it "Y".

There is a third point on p different from X and Y [A_2]; call it "Z". A and Z are not joined [A_5]; P and Z are not joined [D_2]; hence Z is a pole of q [Lemma 2].

But Q is the pole of q [hyp], so that Z = Q [A_3], and we conclude that Q is on p.

T_3: If A, B, and C are joined by l, then the polars a, b, and c are joined by L, which is the pole of l.

Proof: Suppose A, B, and C are joined by l [hyp]. A on l implies a on L [T_2], hence, a on L [mod pon]. Similarly, b on L and c on L. That is, a, b, and c are joined by L [D_1*].

Definition: "Point" and "line" are called *dual* words. That is, we say the dual of "point" is "line", or the dual of "line" is "point". "Pole" and "polar" are also dual words. "On" and "join" are each *self-dual.* If "S" is a statement in the geometry, then its *dual* statement, denoted by "S^*", is obtained from "S" by replacing each undefined or defined word in "S" by its dual.

It follows from the definition that "S^*" is a statement in the geometry, that is, a statement about the terms and relations of the geometry. It is obvious that "$(S^*)^*$" is the same statement as "S".

We have not given the foregoing definition a designation "D_3", because the definition is not a statement *within* the geometry but is a statement *about* the geometry. The statement "A_2", for instance, is a statement about points and lines, whereas the foregoing definition is a statement about "statements of the geometry". The notion of duality is a little like the notion of inference in that it gives a way of deriving new statements from given statements.

Examples of dual statements have been encountered already. "D_1" and "D_1^*" are dual as are "D_2" and "D_2^*". The notion of duality is useful, because we shall show that if a statement in the geometry is true, then so is its dual. To establish this, we first prove the next six theorems.

T_4: (A_0^*) Line p is on point Q if and only if point Q is on line p.

Proof: The statement is just "A_0" with an obvious permutation of names and clauses.

T_5: (A_1^*) There exists at least one point.

Proof: There exists a line [A_1] that has a pole [A_3].

T_6: (A_2^*) There exist exactly three distinct lines on every point.

Proof: Let L be a given point [hyp]. L has a unique polar [A_4], call it "l". l has exactly three distinct points on it [A_2], call them "A", "B", and "C". Now,

$$(\text{A, B, and C on l}) \rightarrow (\text{a, b, and c on L}); \qquad [\text{T}_3]$$

hence,

$$\text{a, b, and c on L.} \qquad [\text{mod pon}]$$

From A_3, it follows that

$$(\text{a and b are not distinct}) \rightarrow (\text{A and B are not distinct});$$

hence,

a and b are distinct. [Contrap inf]

Similarly, a and c, and b and c, are distinct. Thus, we have at least three distinct lines on L.

Suppose there is a fourth line d on L and distinct from a, b, or c. Now,

$$(\text{d on L}) \rightarrow (\text{D on l}), \qquad [T_2]$$

so,

$$\text{D on l.} \qquad [\text{mod pon}]$$

Now,

$$(\text{A} = \text{D}) \rightarrow (\text{a} = \text{d}), \qquad [A_4]$$

but a $\neq$ d; hence,

$$\text{A} \neq \text{D}. \qquad [\text{contrap inf}]$$

Similarly, B $\neq$ D and C $\neq$ D, so that l joins four distinct points [conj inf], which contradicts "A_2".

We conclude that there is not a fourth line d on L.

T_7: (A_3*) Every point has exactly one polar.

T_8: (A_4*) Every line has exactly one pole.

Proof: The proofs of "T_7" and "T_8" are trivial.

T_9: (A_5*) If a line p is not on a point X, but is joined to a line on X by a point M, then p is joined to exactly one other line on X by a point N distinct from M.

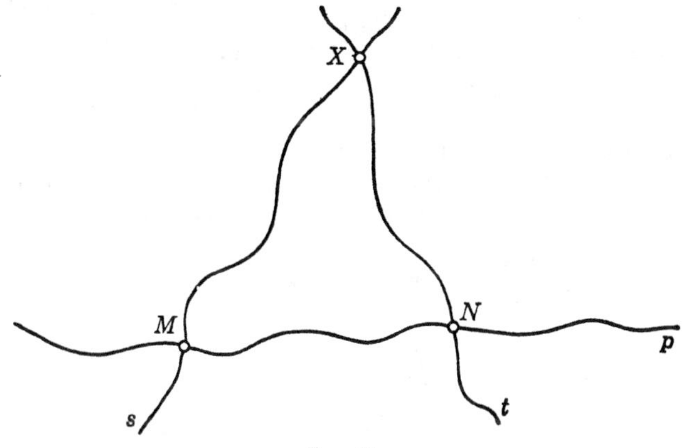

Fig. 19.

Proof: Point X is not on p, but a line on X, call it "s", is joined to p by a point M (Fig. 19) [hyp]. X and M are joined by s [D_1]; hence, X is joined to exactly one other point on p by a line, call it "t", distinct from s [A_5]. Call this point "N", then N joins t and p [D_1*], and t is on X [A_0]. So the theorem is proved.

With the proofs of T_4 to T_9 we have shown that the dual of every axiom is a theorem of the geometry. It is now possible to state and justify the duality principle.

Duality Principle: If "S" is a proved statement in the geometry, then the dual "S^*" can be proved.

Proof: Suppose there exists a sequence of statements,

$$S_1, S_2, \ldots, S_k,$$

that is a demonstration of

$$A_0, \ldots, A_5 \vdash S.$$

Then, consider the sequence of statements

$$S_1^*, S_2^*, \ldots, S_k^*. \tag{4.1}$$

There remains to be shown that each of the "S_i^*" satisfies the requirements of one of the four categories for statements in a demonstration, as given in Sec. 3.8. In other words we must check to see if an "S_i^*" is an axiom, or is in the form of a valid statement formula, or follows from earlier statements by means of an established inference rule, or is a repetition of an earlier statement in the sequence.

Suppose "S_i" is one of the axioms "A_0" to "A_5". Then "S_i^*" is clearly the dual of this axiom.

If "S_i" is in the form of a valid statement formula, then so is "S_i^*".

If "S_i" is the result of an inference from earlier "S's", say

$$\frac{S_f, S_g}{S_i}, \qquad \text{where } f < i \text{ and } g < i,$$

then, clearly,

$$\frac{S_f^*, S_g^*}{S_i^*}$$

is a similar inference.

Finally, if "S_i" is a repetition of some earlier "S_j", then "S_i^*" is a repetition of "S_j^*".

It follows that (4.1) is a demonstration of

$$A_0^*, A_1^*, \ldots, A_5^* \vdash S^*. \tag{4.2}$$

While the proof of "T_4", for example, was not a demonstration of

$$A_0, \ldots, A_5 \vdash A_0^*, \tag{4.3}$$

it gives assurance that such a demonstration exists. Similarly, there exist demonstrations of the remaining duals of the axioms. Thus, with (4.2) and the six statements of which (4.3) is the first, by (3.25) we know that a demonstration of

$$A_0, \ldots, A_5 \vdash S^*$$

exists, and the duality principle is established.

The principle is in effect a special derived inference rule that holds for statements of the geometry. The statement of the duality principle is not itself a statement *within* the geometry but is a statement about statements of the geometry. (Recall that a statement within the geometry is about the undefined or defined words of the geometry.)

We did not mention definition in arguing the duality principle because definitions are just a convenience and not a formal necessity. It is not necessary, for example, to use the word "join". Instead of "A and B are joined by x" we can say "A is on x and B is on x". Similarly, we could dispense with the words "pole" and "polar". Of course, it is impractical to try to get along without definitions, and we shall continue to use them. However, so as not to disturb the use of the duality principle, when a new word is defined, we must take care that the dual of its definition is taken as defining the dual word. This was the case in defining "join" and "pole" and "polar". We proceed to define a new word "triangle".

D_3: A *triangle* consists of three points not on a line and three lines not on a point, with each pair of points joined by one of the lines, and each pair of lines joined by one of the points. The points are called *vertices*. The lines are called *sides*.
Vertex and *side* are dual words.
Triangle is a self-dual word.

That the definition is self-dual can be verified by writing out the dual and comparing the two statements. Further we define:

D_4: A triangle is called *perspective from a point* C if and only if its three vertices are on three lines joined by C, with exactly one vertex on each line.
C is called the *center*.

D_4*: A triangle is called *perspective from a line* c if and only if its three sides are on the three points joined by c, with exactly one side on each point.
c is called the *axis*.

In these definitions "center" and "axis" are dual words. The definitions are duals of each other.

Before proceeding to theorems about triangles, we pause to prove a useful theorem.

T_{10}: If two distinct points are not joined by a line, then each lies on the polar of the other.

Proof: Using the conventions of the pole-polar relation notation, "T_{10}" can be written:

$$(\text{P and Q not joined by a line}) \rightarrow (\text{P on q})(\text{Q on p}).$$

The form of the proof will be that of an indirect proof by cases. Hence, assume

$$(4.4) \qquad (\text{P and Q not joined by a line}) \sim[(\text{P on q})(\text{Q on p})].$$

Now, by the valid form (2.38),

$$(4.5) \qquad \sim[(\text{P on q})(\text{Q on p})] \leftrightarrow [\sim(\text{P on q}) \vee \sim(\text{Q on p})].$$

From (4.4), using conjunctive simplification,

$$\sim[(\text{P on q})(\text{Q on p})],$$

and with (4.5), by modus ponens we conclude

$$\sim(\text{P on q}) \vee \sim(\text{Q on p})$$

is true. We now proceed to the proof that each of the cases leads to a contradiction.

Case I: Assume

$$(\text{P and Q not joined by a line}) \sim(\text{Q on p}).$$

Since P is the pole of p [hyp], Q is not the pole of p [A_3]; so Q is joined to some point on p [D_2], say "R", by a line we shall call "x" (Fig. 20). Now,

P is not joined to R, [D_2]
P is not joined to Q, [hyp]

$$(\text{P not joined to R})(\text{P not joined to Q}) \rightarrow (\text{P is a pole of x})$$

is an instance of Lemma 2. Hence,

P is a pole of x. [conj inf, mod pon]

Then x is the polar of P [Lemma 1], and p is the polar of P [Lemma 1], x is distinct from p, since we have Q on x and ~(Q on p), [hyp of the case]. Thus P has two polars, contradicting "A_4".

Case II: Assume

(P and Q not joined by a line)~(P on q).

A proof that this leads to a contradiction of "A_4" is similar to Case I.

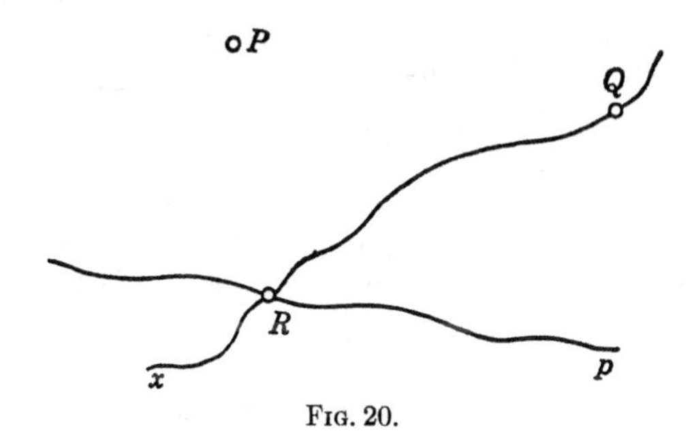

FIG. 20.

With the proof of Case II, the indirect proof by cases is complete. The form of the proof, in outline, was as follows.

To prove:

$$\text{Axioms} \vdash X \rightarrow YZ.$$

In Case I, we proved

$$\text{Axioms}, X{\sim}Y \vdash R{\sim}R.$$

In Case II,

$$\text{Axioms}, X{\sim}Z \vdash R{\sim}R.$$

Hence, by the deduction theorem,

$$\text{Axioms} \vdash X{\sim}Y \rightarrow R{\sim}R.$$
$$\text{Axioms} \vdash X{\sim}Z \rightarrow R{\sim}R.$$

Therefore, by the method of proof by cases, Sec. 3.4,

$$\text{Axioms} \vdash (X{\sim}Y) \vee (X{\sim}Z) \rightarrow R{\sim}R.$$

So we have

$$\text{Axioms} \vdash {\sim}[(X{\sim}Y) \vee (X{\sim}Z)],$$

or by substitution from (2.48),

$$\text{Axioms} \vdash {\sim}[X({\sim}Y \vee {\sim}Z)],$$

and by substitution from De Morgan's law (2.38),

$$\text{Axioms} \vdash \sim[\text{X}\sim(\text{YZ})],$$

or, finally, by substitution in (2.41)

$$\text{Axioms} \vdash \text{X} \rightarrow \text{YZ}.$$

We are now ready to prove a theorem about the existence of triangles. Triangles have been defined, but, of course, a definition never asserts existence. In Euclidean geometry, for example, parallel lines are defined to be lines that do not intersect, but the existence of parallel lines is given in an axiom. It is easy enough to define words that do not name anything existing in the system. For instance, in Euclidean geometry we might agree to call any triangle having an interior angle greater than 300° a "purple triangle". But we could never prove the existence of purple triangles from the axioms of Euclidean geometry.

It would be easy to prove the mere existence of a triangle in the miniature geometry. The next theorem can be used to prove existence of triangles, but asserts a great deal more than mere existence, and consequently has a rather long proof.

T_{11}: Every point has exactly two triangles perspective from it.

NOTE: From now on, it will be convenient to use a new notation for lines. If X and Y are joined by a line, we shall designate this line by the symbol "(XY)". We shall also designate the triangle whose vertices are X_1, X_2, X_3, by the symbol "$(X_1X_2X_3)$".

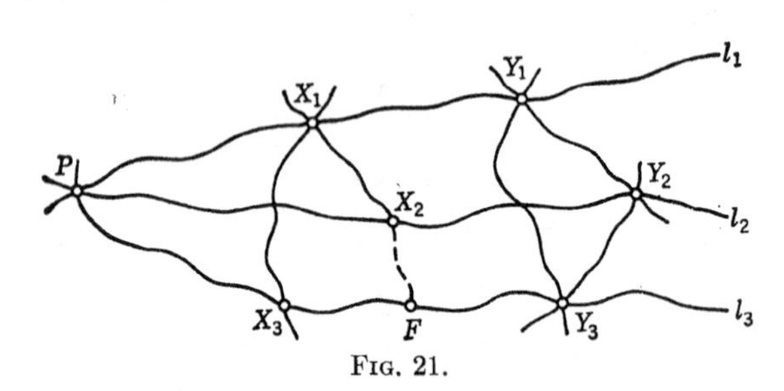

FIG. 21.

Proof: Call the point "P". P joins just three lines [A_2*], call them "l_1", "l_2", and "l_3".

l_1 has exactly two other points on it [A_2]. Call them "X_1" and "Y_1". X_1 is joined to l_2 at P [hyp]; so X_1 is joined to exactly one other point on l_2 [A_5], call it "X_2" (see Fig. 21).

Similarly, X_1 is joined to some point of l_3 other than P. Call it "X_3".

Similarly, X_2 is joined to a point on l_3 different from P, call it "F" (Fig. 21).

We would like to prove that $F = X_3$ so that $(X_1X_2X_3)$ is a triangle. We proceed indirectly.

Suppose $X_3 \neq F$.

Now, the figure suggests that there are three lines joined by X_2. Let us prove that they are indeed all distinct. If (X_1X_2) is the same line as l_2, that is, $(X_1X_2) = l_2$, then X_1 is on l_2 and $l_1 = l_2$ [T_1]. But we have $l_1 \neq l_2$, so we have $(X_1X_2) \neq l_2$ [contrap. inf.].

Similarly, $(X_2F) \neq l_2$.

Finally, if $(X_1X_2) = (X_2F)$, then P is joined to all three points of this line, which contradicts "A_5". Thus, there are three distinct lines (X_1X_2), (X_2F), and l_2 joined by X_2. By "A_2*" these are the only lines on X_2.

Now,

$$(X_2 \text{ joined to P})(X_2 \text{ joined to F}) \rightarrow (X_2 \text{ is not joined to } X_3). \quad [A_5]$$

But

X_2 is joined to P, [hyp, D_1]

X_2 is joined to F. [hyp, D_1]

Therefore,

X_2 is not joined to X_3. [conj inf, mod pon]

Hence, either (X_2P) or (X_2F) or (X_1X_2) is the polar of X_3 [T_{10}], but this is impossible, since we already have that X_3 is joined to a point on each of these lines [D_2*].

The contradiction shows that $X_3 \neq F$ is false, or that $X_3 = F$ is true.

It remains to show that X_1, X_2, and X_3 are vertices of a triangle. Two distinct lines are not joined by more than one point [T_1*], thus, since l_1 and l_3 are joined by P, $X_1 \neq X_3$. Similarly, $X_2 \neq X_3$ and $X_1 \neq X_2$. The method used to show that $(X_1X_2) \neq (X_2F)$ can be applied to show the sides are distinct, so that X_1, X_2, and X_3 are indeed vertices of a triangle.

To get the second triangle, we note that there is a third point on l_1 distinct from P and X_1 [A_2], call it "Y_1". To show that Y_1, Y_2, and Y_3 are vertices of a triangle is exactly similar to the foregoing proof.

That these two triangles are the only triangles perspective from P is left as an exercise.

Corollary: If two triangles are perspective from a point, their vertices and the perspective center form a set of seven distinct points and their sides and the lines on the center form a set of nine distinct lines.

By the duality principle we know that the dual of "T_{11}" is a theorem, namely, "Every line has exactly two triangles perspective from it". The relation between the two types of perspectivity is given in the next theorem. We call it Desargues' theorem because of its similarity to a theorem of that name in Euclidean geometry.

T_{12} (Desargues' theorem): If two triangles are perspective from a point, they are perspective from a line.

Proof: Consider two triangles ($X_1X_2X_3$) and ($Y_1Y_2Y_3$) perspective from a point P (Fig. 22).

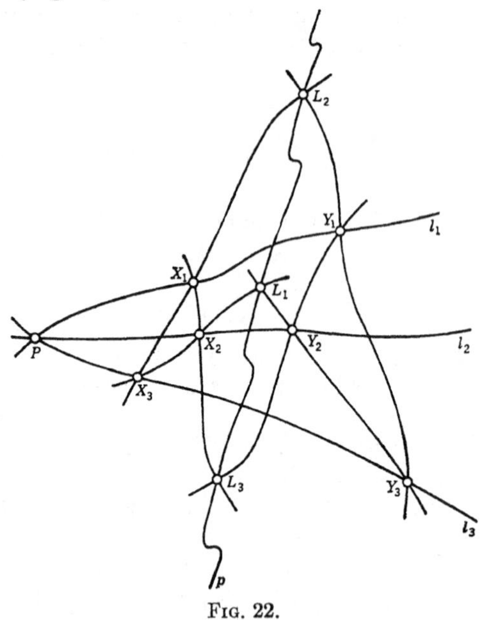

FIG. 22.

Y_3 is joined to P and Y_1 on l_1 [hyp]; hence, Y_3 is not joined to X_1 [indirectly from A_5]. Similarly, Y_3 is not joined to X_2. From an instance of Lemma 2 it follows that

$$(Y_3 \text{ not joined to } X_1)(Y_3 \text{ not joined to } X_2) \rightarrow (Y_3 \text{ is the pole of } (X_1X_2)),$$

therefore,

Y_3 is the pole of (X_1X_2). [conj inf, mod pon]

Or (X_1X_2) is the polar of Y_3 [Lemma 1].

Similarly, (Y_1Y_2) is the polar of X_3. Now,

$$(P, X_3, Y_3 \text{ joined by } l_3) \rightarrow (\text{polars p}, (Y_1Y_2), (X_1X_2) \text{ joined by } L_3), \quad [T_3]$$

where L_3 is the pole of l_3. So,

L_3 joins p, (Y_1Y_2), (X_1X_2). [mod pon]

Similarly, it can be shown that L_1 joins p, (X_2X_3), (Y_2Y_3) and L_2 joins p, (X_1X_3), (Y_1Y_3). It follows that p joins L_1, L_2, L_3 [D_1].

That L_1, L_2, L_3 are distinct is left as an exercise. It is now easily seen that the two triangles are perspective from p.

Corollary 1: The axis of the two perspective triangles is the polar of their center.

D_5: The collection of points and lines consisting of the sides and vertices of two perspective triangles, together with their center and axis and the lines joined by the center, is called a *Desargues configuration.*

Corollary 2: The three lines on any point of a Desargues configuration are lines of the Desargues configuration.

The proof is left as an exercise. Observe that not every pair of triangles need be perspective from a point.

The finiteness of the geometry is proved in the following theorem, which is the last one that we shall prove here.

T_{13}: There exist exactly 10 lines and 10 points.

Proof: The proof is made in terms of the notation of Fig. 22.

There exists a point, call it "P" [A_1*]. P has exactly three lines on it, call them "l_1", "l_2", "l_3" [A_2*], and P has exactly two triangles perspective from it [T_{11}]. Denote the triangles by their vertices as "$(X_1X_2X_3)$" and "$(Y_1Y_2Y_3)$".

The two triangles are perspective from a line, say "p" [T_{12}]. The lines and points so far proved to exist form a Desargues configuration [D_5]. That the points and lines of a Desargues configuration are, respectively, distinct follows from the corollary to "T_{11}" and its dual. It is easy to see that the Desargues configuration contains 10 points and 10 lines. Let us show that there cannot be an eleventh point.

Assume Q is an eleventh point distinct from any point of the Desargues configuration so far constructed. Since every line of the configuration already has three points of the configuration on it [dual of Corollary 2, T_{12}], Q is not on any line of the configuration [A_2].

At least one of the 10 lines of the configuration is not the polar of Q [A_4]; so Q is joined to a point on this line, call it "X". X has three lines of the configuration on it [Corollary 2, T_{12}], and since line (QX) is not one of them, X has four lines on it, which contradicts "A_2*". From the contradiction it follows that there is no eleventh point. By the duality principle, we then know that there is no eleventh line.

While there are more theorems of this miniature geometry, the theorems that we have proved provide a fairly complete description of the properties of the geometry. The proofs presented are, of course, not demonstrations and are still quite lengthy. They are lengthy because we have tried to be careful about proving distinctness of points and lines where this is needed rather than to leave these proofs tacit as is so common in plane geometry proofs. After it happened that we had a line l with points A and B on it, we concluded that there was another point on l distinct from A and B. This is supposed to follow from "A_2". We are dealing here with a little lemma, which may be stated

> A $\neq$ B and A on l and B on l implies there exists C on l, such that A $\neq$ C and B $\neq$ C.

A demonstration of this extremely obvious lemma is not at all short. In fact, before attempting a demonstration, we would find it necessary to restate the axioms in a more suitable form. For instance, "A_2" would have to be restated in a form that spells out more precisely the meaning of the word "exactly" as follows:

> For every line l, there exist three points X, Y, Z, such that X is on l, Y is on l, Z is on l, and X $\neq$ Y, Y $\neq$ Z, Z $\neq$ X, and for all points W, if W is on l, then W = X or W = Y or W = Z.

We have not attempted demonstrations, indeed, we do not yet have the formal machinery to handle demonstrations depending upon such complex statements as the axioms of the miniature geometry. In the restated axiom, the quantifications "For every line l there exist three points X, Y, Z", "for all points W" cannot be adequately handled by the methods of the statement calculus. They depend upon notions to be developed in a later section.

EXERCISES

1. Prove each of the statements that were left as exercises in Sec. 4.2.
- *a.* Page 105: Proof of "B $\rightarrow$ A" of Lemma 1.
- *b.* Page 115: Proof that the two triangles of "T_{11}" are the only triangles perspective from P.
- *c.* Page 117: Proof that L_1, L_2, L_3 of "T_{12}" are distinct.
- *d.* Page 117: Proof of Corollary 2 of "T_{12}".

2. Prove, or disprove, the following conjectures about the miniature geometry:

a. Every triangle has a unique center. See "D_4".

b. If two triangles have no common vertex, then they have the same center.

c. Define lines to be *parallel* if they are not joined by a point. Then consider the conjecture: If P is not on l, then there exists a line on P parallel to l.

d. If P is not on l, then there is not more than one line on P parallel to l.

3. Define *parallel points* in such a way that in view of Exercise 2*c* "parallel" is a self-dual word.

4. Prove or disprove: If p || q, then for their poles, P || Q.

5. The following seven statements are the axioms for a miniature geometry with undefined words *point* and *line* and an undefined relation *on.*

F_0: A point is on a line if and only if the line is on the point.
F_1: There exists at least one line.
F_2: On any two distinct points there exists at least one line.
F_3: On any two distinct points there exists not more than one line.
F_4: Every line has exactly three points on it.
F_5: On any two lines there is at least one common point.
F_6: Not all the points of the geometry are on a single line.

Find and prove some theorems of the geometry. The geometry has a finite number of points and a finite number of lines and has duality. An account of this geometry can be found in Veblen and Young.[1] It is discussed also in a paper by H. F. MacNeish.[2]

4.3 INTERPRETATIONS

In an abstract mathematical system, such as the miniature geometry, the statements of the axioms and the theorems are meaningless statements. They remain meaningless until *interpreted* within some other system, either abstract or real. The miniature geometry, for instance, can be interpreted in the system of arithmetic by means of additional statements, which we can call *correlative definitions,* that tell us what notions of arithmetic are to correspond to the primitive undefined notions of the geometry. We can say that by means of these correlative definitions, the previously undefined notions of the geometry are given *meaning*

[1] *Projective Geometry*, vol. 1, chap. 1, Boston: Ginn and Company, 1918.

[2] "Four Finite Geometries," *The American Mathematical Monthly*, vol. 79, pp. 15–17, January, 1942.

within arithmetic. With these meanings, the axioms of the geometry become statements of arithmetic, and as such are either true or false within arithmetic. If the axioms are all true statements of the arithmetic, then so will all the interpreted theorems of the geometry be true arithmetical statements; the resulting system of statements about arithmetic is called a *valid interpretation* of the geometry. Such a system is also called a *model* of the geometry. The word *model* is, however, often used in exactly the opposite sense. That is, the geometry may sometimes be called a model of what we have called the arithmetic interpretation.

As an example of an interpretation of the miniature geometry, consider the following correlative definitions:

(4.6)

a. The ten points correspond to the ten numbers 0, 1, 2, 3, 4, 5, 6, 7, 8, 9.

b. The ten lines correspond to the ten numbers 256, 578, 047, 679, 289, 019, 038, 123, 146, 345.

c. A point is on a line if and only if the digit corresponding to the point is a digit of the numeral corresponding to the line. If a point and a line have this relation, then the line is also called "on the point".

With the meanings introduced by these correlative definitions, every statement of the geometry becomes a statement about certain numbers. There is, of course, no assurance that the interpreted statements are true statements about numbers. To show that the interpretation is valid, it is necessary to check all the interpretations of the axioms.

It is easy to see that axioms "A_0", "A_1", and "A_2" are satisfied in the interpretation. The definitions "D_1", "D_1^*", "D_2", "D_2^*" can be regarded as defining new words in terms of the interpreted words *point* and *line* and *on*. As an application of "D_1^*", for instance, we would say that lines 047 and 038 are joined by 0.

From "D_1" we know what a pole of a line is, but we do not know whether every (or any) line has a pole, or that each line has a unique pole. This must be checked in order to see if "A_3" is a true statement in the interpretation.

Let us try to find a pole for the line 256. The points 2, 5, and 6 are immediately ruled out as possible poles by the definition. The existence of the line 578 rules out points 7 and 8 as poles for 256, since both 7 and 8 are joined to 5 on 256 by the line 578. In order to make the analysis more systematic we list in column A below some lines and in column B some points. The points in column B are points that cannot be poles of 256 because of the existence of the corresponding line in column A. We do not bother to list in column B points already eliminated.

Column A	*Column B*
256	2, 5, 6
679	7, 9
289	8
123	1, 3
146	4

The only number not listed in column B is 0. To see that 0 is the pole of 256, we observe that the only lines on 0 are 047, 019, and 038, and none of these is on a point of 256. Clearly, we not only have found *a* pole of 256 but *the* pole of 256. In fact, the lines were listed in (4.6) in such a way that 0 is the pole of 256, 1 is the pole of 578, and so on. Once a unique pole for each line is found, we also have a unique polar for each point. It is easy, if tedious, to check that there is just one polar for each point.

Axiom "A_5" is the most onerous to check. We shall check it in the case of point 1 and line 256, and assert without proof that it checks in all cases. Now, 1 is joined to 2 by 123. Also, 1 is joined to 6 by 146. For "A_5" to be satisfied in this case, there must be no line joining 1 and 5, where 5 is the third point on 256. A check of the lines shows that there is no line joining 1 and 5. A complete check could be made similarly, but a little ingenuity can reduce much of the labor involved. We do not carry the matter further here but assert that all the axioms are true statements in the interpretation. Hence, the interpretation is valid.

It follows that all the theorems are true statements in the interpretation. From "T_{11}", "T_{12}", and their duals (pages 114 and 116), for example, every pole-polar pair should determine exactly two perspective triangles. To find the triangles determined by 0 and its polar 256, we first locate the three lines on 0, they are: 047, 019, and 038. If "$\triangle_1$" and "$\triangle_2$" denote the triangles we are looking for, then if 4 on 047 is a vertex of $\triangle_1$, 7 must be a vertex of $\triangle_2$. In the proof of "T_{12}" it was shown that each side of $\triangle_1$ is a polar of a vertex of $\triangle_2$; so 123, which is the polar of 7, must be a side of $\triangle_1$. It follows that $\triangle_1$ has vertices 4, 1, and 3, leaving 7, 9, and 8 for the vertices of $\triangle_2$.

The interpreted lines and points have many other properties and relations besides those given by the axioms. For instance, the point 7 is a prime number, and the line 345 is divisible by the point 3, etc. The additional properties in no way affect the validity of the interpretation.

The given interpretation is abstract, since it is a subsystem of the abstract system of arithmetic. It would be easy to find other arithmetic interpretations different from the one given. In particular, we could interchange the words "point" and "line" in the first two correlative definitions (4.6), and with a suitable rewording of the third definition obtain an interpretation. From the duality principle it is clear that the new interpretation is a valid one.

Another interpretation could be made in which the ten points are ten individuals, and the ten lines are certain committees of three individuals each. "On" could be taken to mean "member of".

More interesting is the following algebraic interpretation.

a. The ten points correspond to the following ordered pairs of integers:

(0, 0)	(5, 0)
(2, −1)	(9, 0)
(4, 2)	(1, 8)
(6, −3)	(6, 5)
(6, 3)	(11, 2)

b. The ten lines correspond to the linear equations:

$$\begin{aligned} 3x + 5y &= 43 & x &= 6 \\ x + y &= 9 & 3x - 2y &= 8 \\ x - y &= 9 & x + 2y &= 0 \\ 2x + y &= 10 & y &= 0 \\ x - 3y &= 5 & x - 2y &= 0 \end{aligned} \tag{4.7}$$

c. A point is on a line (and the line is on the point) if and only if the number pair corresponding to the point satisfies the linear equation corresponding to the line when the first number of the pair is substituted for x and the second for y.

Checking the validity of the interpretation is left as an exercise.

This interpretation is interesting, because it is well known from Descartes' development of analytic geometry that ordered pairs of real numbers and linear equations in two variables can be interpreted in Euclidean geometry with linear equations corresponding to lines and the ordered pairs of real numbers corresponding to points. Then, "point on a line" corresponds to "coordinates of the point satisfy the equation of the line". It follows that the Euclidean points and lines corresponding to the number pairs and linear equations given in (4.7) constitute a Euclidean interpretation of the miniature geometry. As a result, the figure obtained by graphing the equations and number pairs given in (4.7) in the usual way will be of great help in checking the algebraic interpretation given by (4.7). In fact, the streaks and dots of lead or ink on a sheet of paper can be taken as a physical interpretation of the geometry, if appropriate correlative definitions are given.

While it is not our purpose to discuss the scientific method at length in this book, yet the preoccupation with the formal side of mathematics, particularly in these last sections, may tend to give a false picture of mathematics in relation to the real world. We pause a moment to consider the empirical bases for mathematics.

Abstract mathematical systems are not created in a meaningless vacuum independently of an interpretation. It is invariably the case that abstract systems are created with some specific interpretation in mind, which guides the selection of axioms and undefined notions. This is not to say that the guiding interpretation is necessarily a real one, that is, a part of the real world. Yet mathematics has strong empirical roots.[1] The Greek geometry got its start as an empirical science. The profound achievement of Euclidean geometry can be regarded as a step toward abstracting geometry from its empirical roots, but the presence of an interpretation is everywhere felt in the Euclidean formulation, and it cannot stand as a rigorously formulated abstract system independent of its interpretation. It is only in recent times that mathematicians have completed the process of abstraction to produce a rigorous system that can stand independently of an interpretation. Examples are the formulations for Euclidean geometry of Hilbert[2] and Veblen.[3] However rigorous these "de-empiricized" formulations, it is still true that empirical notions provided the inspiration for Euclid's geometry, and hence for these new geometries, since they are just more careful formulations having an interpretation in common with Euclidean geometry. To quote Hilbert, "This problem [the problem of choosing axioms for geometry and investigating their relations] is tantamount to the logical analysis of our intuition of space".[4]

The process of selecting axioms for a system that is to have a preconceived interpretation is inductive. If the interpretation is to be a part of the real world, generally some process of idealization of the real phenomena is necessary, that is, one does not try to deal with his total experience of the phenomena in question, but abstracts certain parts of the experience for consideration. An axiomatic system is sought that will have the idealized phenomena as an interpretation. When such a system is found, it can be examined critically from the point of view of its logical form quite independently of any interpretation. However, since it is the interpretation that is of prime importance, there remains

[1] University of Chicago, *The Works of the Mind*, Chicago: The University of Chicago Press, 1947. See S. Chandrasekhar, "The Scientist," pp. 159–179, and John von Neumann, "The Mathematician," pp. 180–196.

[2] David Hilbert, *The Foundations of Geometry*, La Salle, Ill.: The Open Court Publishing Company, 1902.

[3] Oswald Veblen, "A System of Axioms for Geometry," *Transactions of the American Mathematical Society*, vol. 5, pp. 343–84, 1904.

[4] Hilbert, *op. cit.*, p. 1.

the question of the adequacy of the interpretation as a body of statements about the real world, idealized or not. Criticism from this point of view depends upon careful observation and controlled experimentation.

Successful theoretical systems have tremendous value in systematizing our knowledge of nature and in predicting relationships not previously observed or considered in nature. Logically sound theoretical systems are successful to the extent that they systematize large areas of knowledge, and unsuccessful to the extent that their interpretations differ from a reasonable idealization of the real world. Even though the interpretation of the Newtonian physical theory is now known to differ significantly in places from the observable world, it must be considered a highly successful theory in view of the large area of knowledge it systematizes and its contribution to new knowledge. The new physics is more successful in that it systematizes more and has fewer points of difference between its interpretation and the real world. One can regard the Newtonian interpretation as an approximation to the real world and the new physical interpretation as a closer approximation. It should be noted that the better approximation is earned at the expense of a far more complex theory.

Let us return to the question of precedence, which comes first, the abstraction or the interpretation, the chicken or the egg? It is overwhelmingly the case that an abstract system is developed with some interpretation in mind, but the interpretations for mathematical systems are often themselves abstract. Sometimes the abstractions are so extreme that the empirical roots are hard to find. This is the case with the higher reaches of modern topology and modern algebra. However, in some other newer developments in mathematics, the empirical roots are quite evident, as in probability theory, information theory, and the theory of games.

EXERCISES

1. In the interpretation given by (4.6):

a. Find the two triangles perspective from 9 and 345.

b. Find the pole of 289 and show that "A_3" is satisfied for this pole-polar pair.

c. Find a line parallel to 289 (see Exercise 2*c*, page 119).

2. Write out the correlative definitions necessary to yield an interpretation of the miniature geometry in terms of people and committees, as suggested in the text, page 122.

3. Check the validity of the interpretation given by the correlative definitions in (4.7).

4*a*. Write out the necessary correlative definitions to yield an interpretation of the miniature geometry as a system of pencil streaks (drawn with a straightedge) and pencil dots.

b. Draw the interpretation for these correlative definitions, using the suggestions on page 122.

5. Find an interpretation of the miniature geometry in which points and lines are natural numbers, and where "point x is on line y" means "y ÷ x is a natural number". (HINT: Exploit properties of prime numbers.)

V: THE RESTRICTED PREDICATE CALCULUS

5.1 STATEMENT FUNCTIONS

Our treatment of formal logic up to this point has been limited to consideration of statements, relations between statements, and arguments that can be expressed as sequences of statements. The symbols "A", "B", . . . were always interpreted as translations of whole statements. It was possible to consider some complex statements as compounds of simpler statements, and to express the complex structure symbolically by means of the connectives "&", "v", etc. The machinery so far developed is not yet adequate to reflect all the formal aspects of mathematical arguments. For instance, in the long demonstration of (3.34) the axioms A_1, A_2, A_3, A_4 as written are statements about three natural numbers whose names are "x", "y", and "z", but are treated as statements about *any* three natural numbers. We shall examine this procedure more closely in the succeeding sections. Also, the long demonstration was a proof of

$$a \varepsilon N,\ b \varepsilon N,\ c \varepsilon N,\ A_1,\ \ldots\ ,\ A_4 \vdash (a + b) + c = (b + c) + a,$$

but a theorem about natural numbers ought to be a statement about natural numbers that follows from the axioms, and not from the axioms and some additional *ad hoc* assumptions. We should have liked to deduce from the axioms alone some such statement as

For all natural numbers a, b, c, $(a + b) + c = (b + c) + a$,

but the formal machinery for doing so was lacking. To develop the formal machinery, it will be necessary to examine more closely Steps 3 and 6 of the pattern of conventional proofs given in Sec. 3.12. Step 3 is concerned with stating a geometric theorem in terms of particular, though unknown, points and lines, and Step 6 is concerned with the ultimate generalization from a statement about particular points and lines to a statement about all points and lines.

In the introduction, the Socrates Argument was presented as an example of a formal argument:

All men are mortal.
Socrates is a man.
∴ Socrates is mortal.

In Sec. 2.1 it was asserted that the statement calculus is not adequate to handle such an argument because its validity depends on subject and predicate relations. We could, of course, translate the three statements, respectively, "A", "B", "C", and symbolize the argument

$$\frac{A, B}{C}.$$

But we have no inference rules to apply here to infer "C" from "A" and "B", and so have no formal way of checking the validity of the argument.

To develop the necessary formalism, let us begin by considering the simpler of the two premises, namely,

(5.1) Socrates is a man.

The statement asserts that the individual Socrates has the property of being human (or being a man, or being a member of the class men). The name "Socrates" designates the individual who has the property, and the predicate "is a man" indicates the property. We choose "$H(\ )$" as a symbolic translation of the predicate "is a man", and may then translate (5.1)

$$H(\text{Socrates}).$$

More neatly, we may take "s" as another name for Socrates, and translate (5.1)

(5.2) $$H(s).$$

The symbol "$H(s)$" consists of two parts, the predicate symbol "$H(\ )$" and the individual symbol "s". Statements similar to (5.1) could be translated in a similar way:

(5.3)
$H(e)$: The Eiffel tower is a man.
$H(j)$: Joe is a man.
$H(c)$: This cat is a man.

The first and third statements in (5.3) are false, and the second depends on what individual is designated by "Joe". The statements (5.2) and (5.3) all have a common form, which we symbolize as

$$H(x), \tag{5.4}$$

and read "x is a man". The "x" in (5.4) is an *individual variable* but is not the name *of* any individual, it acts as a place holder for names of individuals. "$H(x)$" is called a *statement function*, and it is *not* a statement. However, "$H(x)$" becomes a statement when the variable is replaced by the name of some individual. By an *individual*, hereafter, we mean not just a human individual, but any individual *thing* such as a particular cat, or Boston, or the number of this page, or the set of all monkeys in Africa. Observe that it makes no sense to talk about the truth value of an interpretation of (5.4), even though we have an interpretation for "H".

A *simple statement function*[1] is defined to be an expression consisting of a predicate symbol and an individual variable, which becomes a statement when the name of an individual is substituted for the variable. The statement resulting from such a substitution is called a *substitution instance* of the statement function.

The statements "$H(s)$", "$H(e)$", "$H(j)$", "$H(c)$" in (5.3) are all substitution instances of "$H(x)$".

Compound statement functions can be built up from simple statement functions by means of the connectives used in the statement calculus. The resulting statement functions will have predicate symbols and individual variables in them, but clearly become compound statements when all the variables are replaced by names of individuals. Let us take in addition to "$H(x)$" the function "$M(x)$" as a translation of

x is mortal.

Then such expressions as follow are compound statement functions:

$$H(x)M(x), \qquad H(x) \rightarrow M(x), \qquad \sim H(x) \vee M(x). \tag{5.5}$$

[1] These are often called *propositional functions*, but we avoid the word since many authors use the word "proposition" to refer to the *meaning* of a statement. However, while we have, and will, interpret "$H(s)$", "$M(s)$", etc., as meaningful sentences, we are interested in displaying a purely formal symbolism.

Substitution instances of (5.5) are compound statements whose truth values depend on the truth values of the corresponding substitution instances of "$H(x)$" and "$M(x)$".

The substitution instances exhibited are called *singular* statements because in each case a single individual is asserted to have the property indicated by the predicate symbols.

EXERCISES

Using the notation of Sec. 5.1 and other appropriate symbols, translate each of the following sentences:

1. Smith is mortal.
2. Jane is mortal.
3. Elita is a man.
4. Joe is a man and Joe is mortal.
5. Buttons is not a man or Buttons is mortal.
6. Don is a man and Don is mortal.
7. If Chimp is a man, then he is mortal.
8. Euclid is both a man and mortal.

With $H(x)$ and $M(x)$ having the interpretations given in Sec. 5.1 translate into English sentences each of the following symbolic statements:

9. $H(\text{cat})$
10. $M(\text{dog})$
11. $H(\text{Joe})M(\text{Joe})$
12. $H(\text{Peter}) \vee M(\text{Peter})$
13. $H(\text{Alice})M(\text{Alice}) \rightarrow M(\text{Alice})$
14. $H(\text{Newton}) \vee \sim M(\text{Newton})$

For each of the following sentences exhibit (1) the individual, (2) the predicate, and (3) a suitable symbolic translation:

15. Joe is an animal.
16. 3 is a prime number.
17. n is a natural number.
18. 2 is even, a natural number, and a prime number.
19. 4 is an even but not a prime number.
20. ABC is a triangle.
21. Triangle ABC is isosceles.
22. A and B are points and l is a line.

5.2 UNIVERSAL QUANTIFIER

Statement functions comparable to "$H(x)$" and "$M(x)$" abound in elementary algebra. If one reflects carefully upon the use of equations (one variable) in elementary algebra, one finds that a good deal of the time they can be regarded as statement functions, rather than as statements. Consider carefully the simple equation

$$(5.6) \qquad 2x + 1 = 7.$$

If the equation (5.6) occurs without any other context, it cannot be regarded as a statement in the sense of preceding sections. One would hesitate to assert that (5.6) is true, or false for that matter, without further information about the symbol "x". Of course, if "x" is known from context to be another name for 7, then (5.6) is just a false statement written in an unnecessarily obscure way. Or, if "x" is another name for 3, then (5.6) is a curious way of writing the true statement

$$2 \cdot 3 + 1 = 7.$$

Part of the power of the algebraic method results from regarding equations such as (5.6) as statement functions having many substitution instances. We may regard (5.6) as expressing the form of all possible statements that can be obtained as substitution instances of it. In algebra, it is customary to call the symbol "x" in (5.6) an *unknown*, and to regard the substitution instance,

$$2 \cdot 3 + 1 = 7,$$

as the result of replacing "x" in (5.6) by "3". In this case 3 is called a *value* of the unknown. However, in algebra, the values of the unknowns are always numbers (reals, or rationals, or integers, etc.). One would not countenance an attempt to replace "x" in (5.6) by the name "Joe", or "Socrates", or "this cat", i.e., these are not considered names of values of "x".

The unknowns of algebra, then, are simply symbols which can take on values, and the values are always numbers of some more or less restricted class. Hence, equations such as (5.6) are simply algebraic expressions involving an unknown.

Quite a different use of a variable "x" is common in elementary mathematics. The expression

$$(5.7) \qquad (x + 1)^2 = x^2 + 2x + 1,$$

may be regarded in two distinctly different ways. We may regard (5.7) in the same way as (5.6), that is, as a statement function that has many substitution instances. On the other hand, (5.7) may sometimes be regarded as a general statement about numbers; (5.7) may be understood to mean

$$(5.8) \qquad \text{For any number } x, (x + 1)^2 = x^2 + 2x + 1.$$

It is important to observe that (5.8) is a statement. It is not a statement function even though it contains the symbol "x". In effect, (5.8) can be interpreted as asserting that every numerical substitution instance of (5.7) is true.

There is nothing about the expression (5.7) that indicates in which of the two senses it is to be regarded. The choice of meaning is determined by context, and generally is easily made by anyone familiar with algebra. Some writers consistently distinguish the two cases by using the ordinary equality sign "$=$" to indicate a statement function, and the symbol "$\equiv$" to indicate the general statement. Thus, they might write (5.8)

$$(x + 1)^2 \equiv x^2 + 2x + 1.$$

These distinct uses of variables occur in all branches of mathematics. Many students make the distinction consciously for the first time in trigonometry. The equation

$$\sin \theta + \cos \theta = 1 \tag{5.9}$$

is clearly not true for all substitution instances. Such an equation generally appears in a problem context. The problem usually is to find all true substitution instances of (5.9), or in the language of trigonometry, to find all values of θ for which (5.9) is true.

The equation

$$\sin^2 \theta + \cos^2 \theta = 1, \tag{5.10}$$

generally appears in a *proof* context. The problem is to prove that every substitution instance of (5.10) is true, or, in the language of trigonometry, to prove that (5.10) is an identity.

The beginning trigonometry student is not always clear on this distinction and may be baffled by the restrictions or formal rules he is required to observe in proving identities. It seems persuasive to him to prove the identity

$$\frac{\cos \theta}{1 - \sin \theta} = \frac{1 + \sin \theta}{\cos \theta}, \qquad \sin \theta \neq 1, \qquad \cos \theta \neq 0, \tag{5.11}$$

by merely clearing the equality of fractions to obtain

$$\cos^2 \theta = 1 - \sin^2 \theta,$$

which he recognizes as a form of a well-known identity. After all, this is what he would do if he were to treat (5.11) as a trigonometric equation whose solution is desired.

It is important in logic to be perfectly clear about the sense in which variables are used, and it is desirable that the notation indicate the sense. The symbols "$=$" and "$\equiv$" serve well enough to distinguish equations from statements of identity, but are not adequate in the case of a statement function that does not involve equality. What is needed is some symbol that can be written next to a statement function and whose presence indicates that the property symbolized by the predicate holds for all individuals in the universe. Standard symbols used for this purpose are

$$\forall(x) \text{ and } (x),$$

and they always precede the statement function to which they are attached. Either of these symbols is called a *universal quantifier*. In what follows we shall always use the second and simpler symbol "(x)" to denote a universal quantifier. Strictly, the quantification symbol is "( )"; then whatever variable occurs within these parentheses is the one quantified. We may quantify the statement functions

$H(x)$: x is a man.
$M(x)$: x is mortal.

and obtain the symbolic statements

$$(x)H(x), \qquad (x)M(x).$$

The first, "$(x)H(x)$", may be read variously

For every x, x is a man.
For all x, x is a man.
Everything is a man.
Every individual in the universe is a man.

If (5.7) is quantified in the same way, one gets

$$(x)[(x + 1)^2 = x^2 + 2x + 1]$$

which could be read

(5.12) For every possible thing x in the universe, $(x + 1)^2 = x^2 + 2x + 1$.

The statement (5.12) would have the same meaning as (5.8) if the universe of values for x is a universe solely of numbers. In algebra, the variables used generally have numbers for values so it might be reasonable to restrict the universe of objects to complex numbers.

However, in geometry we might want to make statements about "all triangles" or "all points" or "all circles", etc., so that a restriction on the universe of values for variables presents some difficulties. If the universe is to contain such a variety of objects as "points", "triangles", etc., it is not easy to imagine a simple quantified statement that one would like to interpret as true. However, we can employ the logical connectives

"&", "v", "$\rightarrow$" of the statement calculus in a new way to obtain compound statement functions that can be quantified to yield true statements.

Recall that the conditional symbol "$\rightarrow$" is used to form the statement "$A \rightarrow B$" from the statements "A" and "B". We now use "$\rightarrow$" to form a new compound statement function,

$$H(x) \rightarrow M(x), \tag{5.13}$$

from the statement functions "$H(x)$" and "$M(x)$". It may be read variously

If x is a man, then x is mortal.
x is a man implies x is mortal.
x is a man only if x is mortal.

Of course, (5.13) is not a statement, and it makes no sense to interpret it as true or false. If "x" is replaced by "Socrates" in (5.13) then we have

$$H(\text{Socrates}) \rightarrow M(\text{Socrates}), \tag{5.14}$$

which we read: If Socrates is a man, Socrates is mortal. So (5.14) is just a compound statement formed from two simple statements by means of the conditional connective of the statement calculus. We shall call (5.13) a *conditional statement function.*

In an analogous manner we shall freely use the other connectives of the statement calculus to form compound statement functions such as

$$A(x) \vee B(x),$$
$$A(x)B(x),$$
$$A(x) \leftrightarrow B(x),$$
$$\sim[A(x) \vee B(x)] \leftrightarrow \sim A(x) \sim B(x),$$

and so on.

If the universal quantifier is attached to (5.13) we get

$$(x)[H(x) \rightarrow M(x)]. \tag{5.15}$$

Various translations of (5.15) are

(5.16)
For every x, if x is a man then x is mortal.
Everything which is a man is mortal.
All men are mortal.

It is not uncomfortable to regard any one of the readings (5.16) as true, but we cannot use truth-table techniques on (5.15) in an attempt to find the relation between the truth of (5.15) and the truths of its components, since the components are not statements but statement functions and a quantifier. To put it another way, (5.15) might be a perfectly legitimate statement of our statement calculus with symbolic translation "A", say, but "A" cannot be expressed as any compound of simpler statements by the methods of the statement calculus. If there were just a finite number of substitution instances for (5.13), then each could be checked and if they all were interpreted as true, then (5.15) could be interpreted as true. However, useful interpretations of the predicate calculus usually involve infinite universes, and so this method of checking truth is not available. Hence, if we wish to attach notions of truth or falsity to a quantified statement such as (5.15), we must simply attach the label T or F to (5.15) as a whole, and not attempt to attach these labels to its components.

Enough machinery has now been developed to translate the Socrates argument symbolically. With the translations

s: Socrates
$H(x)$: x is a man.
$M(x)$: x is mortal.

the argument is translated

$(x)[H(x) \rightarrow M(x)]$	All men are mortal.
$H(s)$	Socrates is a man.
$\therefore M(s)$	$\therefore$ Socrates is mortal.

Eventually we shall want to have inference rules for quantified statements which will permit determination of the validity of such an argument by formal means. Before these rules are developed we wish to develop a symbolism for translating particular statements of the form "Some_____is_____".

EXERCISES

Translate each of the following compound statement functions into a sentence. Give three different readings of each sentence.

1. $(x)M(x)$
2. $(y)[H(y) \vee M(y)]$
3. $(x)[H(x) \leftrightarrow M(x)]$
4. $(x)[H(x)M(x)]$
5. $(x)[M(x) \rightarrow H(x)]$
6. $(x)\sim[H(x) \vee M(x)]$

Set up appropriate symbolic translations of statement functions, and use them to translate each of the following sentences:

7. Everything which is a number is a complex number.
8. All real numbers are complex numbers.
9. All natural numbers are either odd or even.

10. All mathematicians are logicians.
11. Everything which is a number is imaginary or real.
12. All triangles are isosceles.
13. All equilateral triangles are isosceles.
14. All squares, rectangles, and parallelograms are quadrilaterals.

5.3 EXISTENTIAL QUANTIFIER

The following statements may be considered to have the same meaning.

There exists a man.
Something is a man.
There is at least one x such that x is a man.
There exists an x such that x is a man.

The last of these statements can be partially symbolized:

(5.17) There exists an x such that $H(x)$.

Truth of (5.17) could be interpreted as meaning that there is at least one true substitution instance of "$H(x)$". Using a symbol "$(\exists\quad)$", called an *existential quantifier*, we may write (5.17)

$$(\exists x)H(x).$$

The statement, "Some men are clever",[1] may be variously paraphrased:

Something is a man and clever.
There is at least one thing that is a man and is clever.
There exists an x such that x is a man and x is clever.

The last statement is easily symbolized:

$$(\exists x)[H(x)C(x)].$$

With the universal quantifier, the existential quantifier, and statement functions, it is now possible to translate all the Aristotelian propositions, and hence any Aristotelian syllogism, though as yet we have no formal inference rules for determining validity of such syllogisms.

(5.18)

A.	All men are clever	$(x)[H(x) \rightarrow C(x)]$.
E.	No men are clever	$(x)[H(x) \rightarrow \sim C(x)]$.
I.	Some men are clever	$(\exists x)[H(x)C(x)]$.
O.	Some men are not clever	$(\exists x)[H(x)\sim C(x)]$.

[1] There are those who would interpret "some" here to mean "at least two". Such a use of "some" is not at all convenient in mathematics, and we shall not use it that way in what follows.

It is well to point out that statements (5.18) are taken in the modern sense, that is, (5.18A) and (5.18E) are not statements of existence, as they would be in Aristotelian logic. While (5.18A) can be interpreted as meaning that every substitution instance of

$$H(x) \rightarrow C(x) \tag{5.19}$$

is true, nothing is asserted about existence of true substitution instances of "$H(x)$" or of "$C(x)$". For example, if "$H(a)$" is a false substitution instance of "$H(x)$", then "$H(a) \rightarrow C(a)$" is a true substitution instance of (5.19). See (2.3). Indeed, if the class of men is empty, then every substitution instance of "$H(x)$" is false, and every substitution instance of (5.19) is true, so that in this case (5.18A) would be interpreted as a true statement.

Statements (5.18I) and (5.18O) are clear statements of existence. If either "$H(x)$" or "$C(x)$" has no true substitution instances, then (5.18I) would be interpreted as false.[1]

EXERCISES

Devise suitable symbolic translations of statement functions, and use them to translate each of the following statements:

1. There exists an x such that x is a man.
2. There exists an x such that x is mortal.
3. Something is a man and mortal.
4. Some men are clever and mortal.
5. There is at least one thing that is a man or is clever.
6. Some nails are sharp.
7. Some nails are not sharp.
8. All nails are sharp.
9. No nails are sharp.

Translate the following statements into symbolic translations using the statement functions with their translations as given below:

$\mathrm{Re}(x)$: x is a real number	$P(x)$: x is a prime number
$C(x)$: x is a complex number	$O(x)$: x is an odd number
$\mathrm{Ra}(x)$: x is a rational number	$E(x)$: x is an even number

10. All real numbers are complex numbers.
11. Some real numbers are not rational.
12. There exists a real number that is neither odd nor even.

[1] We rule out the trivial logic in which the universe of values for a variable is empty. Hence, for any statement function "$F(x)$" there is always at least one substitution instance. This does not mean, of course, that "$F(x)$" necessarily has any true substitution instances.

13. There is something that is a prime number.
14. All prime numbers are rational.
15. Some complex numbers are even.
16. Some real numbers are not prime numbers.
17. Not all real numbers are rational.
18. Numbers that are odd or even are rational.
19. There exist both odd and even prime numbers.
20. There exists a prime number and it is odd or even.
21. All prime numbers are odd or even.

5.4 TRANSFORMATIONS OF QUANTIFIERS

For any statement function "$W(x)$", it seems reasonable that

$$\text{"}\sim[(x)W(x)]\text{" is the same as "}(\exists x)[\sim W(x)]\text{".}$$

If "$W(x)$" is a translation for "x is white", then we would naturally interpret "Not, everything is white" and "Something is not white" to have the same meaning.

It is revealing to interpret the foregoing relation in a finite universe of values. Suppose we are dealing with a universe containing just three distinct objects named "a", "b", "c", so that the statement function "$W(x)$" becomes a statement when and only when one of these names is substituted for "x". Then it is reasonable to interpret "$(x)W(x)$" as meaning the conjunction

$$W(a)W(b)W(c).$$

Such a meaning is consistent with the interpretation that "$(x)W(x)$" is true if and only if every substitution instance of it is true. Then "$\sim[(x)W(x)]$" means

$$\sim[W(a)W(b)W(c)], \tag{5.20}$$

and by applications of the De Morgan law (2.38), (5.20) is equivalent to

$$\sim W(a) \vee \sim W(b) \vee \sim W(c). \tag{5.21}$$

The disjunction (5.21) is true exactly when at least one of its parts is a true statement, that is, if there exists at least one true substitution instance of "$\sim W(x)$". But this is exactly the way we have interpreted the truth of "$(\exists x)[\sim W(x)]$".

With an infinite universe of values, the foregoing argument cannot be used. Nevertheless the interpretation is reasonable, and we make the following formal assumption:

$$\sim[(x)W(x)] \leftrightarrow (\exists x)\sim W(x). \tag{5.22}$$

By obvious substitution it is possible to derive the additional equivalences

$$(5.23)\qquad \begin{aligned} (x)W(x) &\leftrightarrow \sim[(\exists x)\sim W(x)], \\ \sim[(\exists x)W(x)] &\leftrightarrow (x)\sim W(x), \\ (\exists x)W(x) &\leftrightarrow \sim[(x)\sim W(x)]. \end{aligned}$$

We can eliminate the square brackets in (5.22) and (5.23) without any danger of confusion. It is only necessary to keep in mind that a negation sign never negates the quantifier, but always the whole quantified statement. Rewritten in this way, (5.22) and (5.23) become

$$(5.24)\qquad \begin{aligned} &a.\ \sim(x)W(x) \leftrightarrow (\exists x)\sim W(x), \\ &b.\ (x)W(x) \leftrightarrow \sim(\exists x)\sim W(x), \\ &c.\ \sim(\exists x)W(x) \leftrightarrow (x)\sim W(x), \\ &d.\ (\exists x)W(x) \leftrightarrow \sim(x)\sim W(x). \end{aligned}$$

EXERCISES

In each case indicate symbolic translations of appropriate statement functions, and use them to translate the given statement:

1. Every rational number is real.
2. There is a number that is rational.
3. Some numbers are not rational.
4. No real number has a square that is negative.
5. The equation $x^2 + 1 = 0$ has no real solution.
6. The equation $x^3 + x^2 + x + 1 = 0$ has a real solution.
7. There is at least one man who if renominated can be elected.
8. If someone is nominated, then someone can be elected.
9. Any polygon has an inside and an outside.
10. There are not any polygons having only two sides.

Use the statement functions translated below, and write an interpretation of each of the following statements:

$I(x)$: x is integral	$P(x)$: x is periodic in decimal form
$N(x)$: x is a number	$T(x)$: x terminates in decimal form
$\mathrm{Ra}(x)$: x is rational	$O(x)$: x is odd
$\mathrm{Re}(x)$: x is real	$C(x)$: x is complex

11. $P(1/3)$
12. $T(1/4)$
13. $\sim\mathrm{Ra}(\sqrt{2})$
14. $\sim\mathrm{Ra}(\sqrt{2}) \to (\exists x)[N(x)\mathrm{Ra}(x)]$
15. $(x)[N(x) \to (\mathrm{Ra}(x) \to P(x) \vee T(x))]$
16. $(x)[N(x)\mathrm{Re}(x) \to (P(x) \vee T(x) \vee \sim\mathrm{Ra}(x))]$

17. $(x)[N(x)\mathrm{Ra}(x) \rightarrow \mathrm{Re}(x)]$
18. $(x)[N(x)I(x) \rightarrow (0(x) \rightarrow 0(x^2))]$
19. $(x)[N(x)I(x) \rightarrow (0(x^2) \rightarrow 0(x))]$
20. $(\exists x)[N(x)C(x){\sim}\mathrm{Ra}(x)]$

5.5 FREE AND BOUND VARIABLES

In any formula, a part of the form "$(x)F(x)$" or "$(\exists x)F(x)$" is called an *x-bound* part of the formula. The *scope* of the quantifier in either case is just the statement function "$F(x)$".

Any occurrence of "x" in an x-bound part is called a *bound occurrence.* Any occurrence of "x" that is not a bound occurrence is called a *free occurrence.*

A statement function lies within the scope of "(x)" [or "$(\exists x)$"] if it lies immediately to the right of "(x)" or if it is a component of some compound statement function contained in parentheses (or brackets) immediately to the right of "(x)".

The quantifier and its scope are included in the boxes in the following examples, i.e., the boxes indicate the x-bound parts:

a. $\boxed{(x)A(x)}$

b. $\boxed{(x){\sim}A(x)}$

c. $\boxed{(x)A(x)} \vee B(x)$

d. $\boxed{(x)[A(x) \vee B(x)]}$

e. $\boxed{(\exists x)A(x)} \rightarrow B(x)$

f. $\boxed{(\exists x)[A(x) \rightarrow B(x)]}$

g. $\boxed{(x)[{\sim}A(x)B(y)]}$

h. $\boxed{(\exists x)A(x)}\,B(x)$

In the foregoing examples, *a, b, d, f* are symbolic statements; in each case the variable is bound by the quantifier. The expressions *c, e, g, h* are statement functions; each contains a free variable. In *c*, for example, "$B(x)$" is not in the scope of "(x)", so that the occurrence of "x" in "$B(x)$" is free, even though the occurrence of "x" in "$A(x)$" is bound.

It might be less confusing to write c:

$$(y)A(y) \vee B(x)$$

so as to avoid any possibility of mistaking the free and bound variables.

The function "$B(y)$" in g lies in the scope of the quantifier, but in this case "y" is free because it is "x" that is quantified. Examples e and h are analogous to c in that "x" is bound in one part and free in another. In each case it would be permissible to choose a symbol different from "x" for the bound occurrences.

It is well to observe that when "$A(x)$" is a given statement function, "$(x)A(x)$" is a statement, and does not have substitution instances. Neither "$(\exists x)$" nor "(x)" is a statement function, and we have no interpretation for substitution of the name of an object for "x" in either quantifier. Hence, to obtain a substitution instance of a statement function such as

$$(x)A(x) \rightarrow B(x), \tag{5.25}$$

one substitutes the name of some object for the *free* occurrence only. If "s" is such a name, then a substitution instance of (5.25) is

$$(x)A(x) \rightarrow B(s),$$

which can be interpreted

If all objects x of the universe have property A,
then object s has property B.

For the same object "s", c, e, g, h of the example have for substitution instances the statements

$$(x)A(x) \vee B(s),$$
$$(\exists x)A(x) \rightarrow B(s),$$
$$(x)[\sim A(x)B(s)],$$
$$(\exists x)A(x)B(s).$$

Upon occasion later on expressions like "$(x)P$" will appear where "x" is not a variable, bound or free, in "P". If, for instance, "P" is a translation for

$$2 + 3 = 5, \tag{5.26}$$

then "$(x)P$" may be translated

(5.27) For everything in the universe, $2 + 3 = 5$.

The word "everything" in the quantifying phrase does not refer to anything at all in the statement "$2 + 3 = 5$"; hence, it is reasonable to take (5.26) and (5.27) as having the same meaning. Such vacuous quantification produces oddly redundant statements, but does no harm and is formally useful.

EXERCISES

List the bound and free occurrences in each of the following examples:

1. $(x)\mathrm{Re}(x)$
2. $(x)[\mathrm{Re}(x) \rightarrow C(x)]$
3. $(x)T(x) \rightarrow \mathrm{Ra}(x)$
4. $R(x) \rightarrow (\exists x)P(x)$
5. $(\exists x)A(x){\sim}B(x)$
6. $(y)(\exists x)[P(x)Q(y)] \rightarrow R(x)T(y)$
7. For Exercises 1 to 6 inclusive indicate which are statements and which are statement functions.

Use the statement functions translated below, and write a grammatical interpretation of each of the following statements:

$Q(x)$: x is a quadrilateral.
$P(x)$: x is a parallelogram.
$Pa(x)$: x has opposite sides parallel.

8. $(x)[Q(x)Pa(x) \rightarrow P(x)]$
9. $(\exists x)[Q(x)P(x)]$
10. $(x)[P(x) \rightarrow Q(x)]$
11. $(\exists x)[Q(x){\sim}P(x)]$
12. $(x)[Q(x) \rightarrow P(x) \vee {\sim}P(x)]$
13. ${\sim}(x)[Q(x){\sim}P(x)]$
14. ${\sim}(x)[Q(x)Pa(x)]$

Indicate symbolic translations of appropriate statement functions, and translate each of the following statements:

15. For every x, if x is a square, x is a rectangle.
16. There is an x such that x is a regular polygon.
17. For all θ, θ is a real number and $\sin^2 \theta + \cos^2 \theta = 1$.
18. For all a, b, c; a, b, c are complex numbers and $a(b + c) = ab + ac$.
19. For some q, q is a quadrilateral and a trapezoid.
20. For some p, p is a polygon and p can be inscribed in a circle.
21. There exists a p such that p is a regular polygon and p can be inscribed in a circle with use of straightedge and compass.
22. For all n, n is a number and n is real or imaginary.

5.6 INFERENCE RULES FOR QUANTIFIED STATEMENTS

There is no necessity for adding to the inference rules of the statement calculus to accommodate quantified statements as long as quantified statements are treated as indivisible units in the logic. However, without rules of inference that permit making use of the internal structure of quantified statements, it is not possible to prove the validity of the simplest syllogism. It would not be possible to prove the validity of the Socrates Argument

$$\begin{array}{l} (x)[H(x) \to M(x)] \\ H(s) \\ \therefore\ M(s) \end{array}$$

or to prove validity for such an obvious argument as

$$\begin{array}{l} (x)A(x) \\ (x)[A(x) \to B(x)] \\ \therefore\ (x)B(x). \end{array}$$

Rules are needed to indicate how to infer:
a. Statement functions from quantified statements.
b. Statement functions from statement functions.
c. Quantified statements from statement functions.
d. Singular statements from quantified statements and vice versa.
In addition to new inference rules, it will be necessary to restate the definition of a demonstration.

5.7 INFERENCE OF A STATEMENT FUNCTION FROM A GENERAL STATEMENT

The Socrates Argument

All men are mortal.
Socrates is a man.
∴ Socrates is mortal.

has been given the symbolic translation

$$\begin{array}{l} (x)[H(x) \to M(x)] \\ H(s) \\ \therefore\ M(s). \end{array}$$

To demonstrate validity means to exhibit a demonstration of

(5.28) $$(x)[H(x) \to M(x)],\ H(s) \vdash M(s).$$

We start out:

1. $(x)[H(x) \to M(x)]$	1. Hyp
2. $H(s)$	2. Hyp

At this point we have no formal way of proceeding according to methods so far introduced. We should like Step 3 to be

3. $H(s) \to M(s)$ 3. ?

In fact, since we interpret "$(x)[H(x) \to M(x)]$" to be true if and only if all substitution instances of "$H(x) \to M(x)$" are true, the third statement seems a reasonable inference from the first. Then the demonstration concludes:

4. $M(s)$ 4. mod pon [2,3]

All that is needed in the formalism is an explicitly stated inference rule that permits the inference of Step 3 from Step 1. We might formulate such a rule as follows:

(5.29) For any statement function "$F(x)$",

$$\frac{(x)F(x)}{F(s)}$$

where s = Socrates.

The inference rule (5.29) is adequate for the demonstration of (5.28), but is a bit restricted. The rule would be as reasonable if we used "Smith" or "Robinson" or any other name in place of "Socrates". We restate (5.29), using a free variable "y" in place of "s".

(5.30) Inference Rule IU: For any statement function "$F(x)$",

$$\frac{(x)F(x)}{F(y)}.$$

In this inference rule, "y" is conceived as representing the name of any arbitrary object in the universe, known or unknown. The inference rule (5.30) will be referred to by the combination of capital letters, IU, and is interpreted as an *inference of a statement function from a universal statement.* Hitherto our inference rules have involved only statements, but (5.30) may have a singular statement or a statement function as its conclusion.

5.8 INFERENCE RULES OF THE STATEMENT CALCULUS EXTENDED

Since the Socrates Argument appears to be valid even if some other name replaces "Socrates," it seems reasonable to try to construct a demonstration proving

$$(x)[H(x) \to M(x)],\ H(y) \vdash M(y)$$

where "y" is an unknown. The demonstration proceeds as before until Step 4 is reached.

1. $(x)[H(x) \to M(x)]$	1. Hyp
2. $H(y)$	2. Hyp
3. $H(y) \to M(y)$	3. IU
4. $M(y)$	4. ?

Certainly "$M(y)$" is not obtained by modus ponens as that rule was used in the statement calculus, since modus ponens is a rule for inferring a statement from given statements. What we would like is an inference rule such as

$$\frac{A(y) \to B(y),\ A(y)}{B(y)}. \tag{5.31}$$

Is such an inference scheme reasonable? Furthermore, since we cannot talk about the truth of the components of the inference, how can it be interpreted?

The inference is justified by considering what happens when we take any substitution instance of the functions in (5.31). If "y" is replaced by the name of some object, then the three components of (5.31) are all statements, and (5.31) becomes just a case of the old modus ponens.

Let us agree to enlarge the concept of modus ponens to include the case (5.31) and retain the designation "modus ponens" to cover both the old inference involving statements and the new inference involving statement functions.

All the other inference rules of the statement calculus will be likewise extended to include statement functions as components. For instance,

$$\frac{A(x),\ B(x)}{A(x)B(x)} \qquad \text{conj inf}$$

$$\frac{P(x) \to Q(x),\ \sim Q(x)}{\sim P(x)} \qquad \text{contrap inf}$$

5.9 EXTENSION OF THE NOTION OF VALID STATEMENT FORMULA

The valid statement formulas of the statement calculus were useful in demonstrations and in providing new derived inference rules. Their validity is easily established by the method of truth tables. Consider the expression

$$\sim[A(x) \vee B(x)] \leftrightarrow \sim A(x) \sim B(x). \tag{5.32}$$

While (5.32) is not a statement formula, it is a statement function having a free variable "x". Any substitution instance of it is a true statement in the form of one of the De Morgan laws [see (2.39)].

In the same way, any other expression involving statement functions, which is in the form of a valid statement formula will have only true substitution instances. It seems reasonable to extend the notion of *valid statement formula* to include cases involving statement functions. We shall retain the phrase "valid statement formula" for both the old and the extended cases. In the following examples, "a" is understood to be the name of some known object. In all cases, the substitution instances are statements in the form of some statement formula that was proved to be valid in an earlier section.

Examples

a. $\sim[A(x)B(x)] \leftrightarrow \sim A(x) \vee \sim B(x)$
b. $[A(x) \rightarrow B(x)] \leftrightarrow \sim A(x) \vee B(x)$
c. $[A(x) \rightarrow B(y)] \leftrightarrow \sim A(x) \vee B(y)$
d. $[A(a) \rightarrow B(y)] \leftrightarrow \sim A(a) \vee B(y)$
e. $[P(x) \rightarrow Q(x)][Q(x) \rightarrow R(x)] \rightarrow [P(x) \rightarrow R(x)]$
f. $P(x)Q(y) \rightarrow [P(x) \rightarrow Q(y)]$
g. $P(a)Q(y) \rightarrow [P(a) \rightarrow Q(y)]$
h. $P(x)Q(y) \rightarrow P(x)$
i. $P(a)Q(y) \rightarrow P(a)$
j. $P(x)Q(a) \rightarrow P(x)$

5.10 THE GENERALIZATION PRINCIPLE

The axioms of mathematics and the theorems that follow from them are statements rather than statement functions. When the axioms and theorems are expressed symbolically, all variables are bound by quantifiers. Theorems of plane geometry are almost always general statements, although the conventional modes of stating them may obscure the quantification. For example, the isosceles triangle theorem is conventionally stated

> If a triangle is isosceles, then the angles
> opposite the equal sides are equal.

This is clearly a statement about all triangles that are isosceles, and a symbolic translation would start

$$(x)[(x \text{ is a triangle})(x \text{ is isosceles}) \rightarrow \underline{\qquad}].$$

In demonstrating this theorem, the mathematician does not seek a sequence of quantified statements, but rather starts with an arbitrary triangle "ABC" in which "AC = BC", and deduces "$\angle$A = $\angle$B". During the proof, "ABC" is viewed as an unknown, but fixed, triangle.

It would ruin the proof to allow triangle "ABC" to vary from step to step in the proof. By the time "∠A = ∠B" is reached, there exists a demonstration of

Axioms of geometry, ABC is a △, ABC has
AC = BC ⊢ ABC has ∠A = ∠B.

The expressions "ABC is a △", "ABC has AC = BC", "ABC has ∠A = ∠B", are extremely abbreviated forms of statement functions. Applying the deduction principle[1] yields first

Axioms, ABC is a △ ⊢ (ABC has AC = BC) → (ABC has ∠A = ∠B),

and applying the deduction principle again,

(5.33) Axioms ⊢ (ABC is a △) → [(ABC has AC = BC) → (ABC has ∠A = ∠B)].

If the valid statement formula (2.55) is used in the extended sense, (5.33) may be rewritten

(5.34) Axioms ⊢ (ABC is a △)(ABC has AC = BC) → (ABC has ∠A = ∠B).

At this point the mathematician feels free to write, without further ado,

(5.35) Axioms ⊢ (ABC)[(ABC is a △)(ABC has AC = BC) → (ABC has ∠A = ∠B)].

As justification for this step, one can say that while throughout the demonstration of (5.34), "ABC" is conceived as naming a fixed object, it is a perfectly arbitrary object, nothing is needed for the demonstration beyond this.[2] To put it another way, the demonstration of (5.34) goes through in the same way whatever object ABC may be.

The step from (5.34) to (5.35) can be stated in general form as the *Principle of Generalization to a Universal Statement.* We shall refer to this principle in what follows by the initials PGU*.[3]

(5.36) PGU*: For any statement function $F(x)$, if $A_1, A_2, \ldots, A_m \vdash F(x)$, then $A_1, A_2, \ldots, A_m \vdash (x)F(x)$, where each of the A's may be a statement or statement function, but x is not a free variable in any of the A's.

[1] As stated in Sec. 3.11, the deduction principle applies to statements, but it can be proved to hold in the extended sense for statement functions as well.

[2] Observe that if ABC is not a triangle then the expression to the right of ⊢ in (5.34) is surely true since it is a conditional with a false antecedent. It is the same situation if ABC is a triangle but not isosceles.

[3] In an axiomatic development of the predicate calculus, PGU* is a theorem. Proofs can be found in Church, Rosser, and others.

We refer to the principle (5.36) as "PGU*" with the asterisk since in later developments (Sec. 5.11) we shall find it necessary to restrict the use of this principle further.

It is extremely important when using PGU* to observe the restriction that "x" not occur as a free variable on the left side of the turnstile. Failure to observe this restriction can lead to some mighty queer results, as in the following example from solid geometry:

(5.37) **Theorem:** All lines perpendicular to a given line b at a point A lie in the plane π that is perpendicular to b at A.

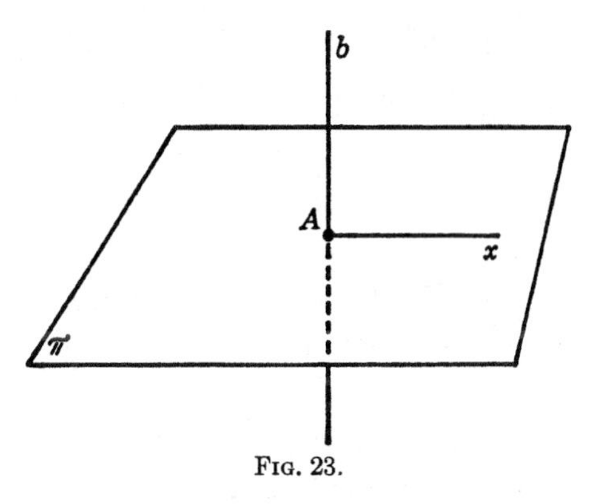

Fig. 23.

Let us make the following translations (Fig. 23):

$$P: \pi \perp \text{b at A.}$$
$$L(x): x \text{ is a line.}$$
$$H(x): x \perp \text{b at A.}$$

Then the normal way to prove (5.37) would be to demonstrate

(5.38) $$P, L(x), H(x) \vdash x \text{ lies in } \pi.$$

If we incorrectly apply the generalization principle PGU* at this stage, we get

$$P, L(x), H(x) \vdash (x)(x \text{ lies in } \pi),$$

which may be translated

$\pi \perp$ b at A and x is a line and $x \perp$ b at A yields, everything lies in π.

This absurd statement results because we tried to generalize on "x" from assumptions "$L(x)$" and "$H(x)$" having "x" free.

Before generalizing, we have to eliminate the functions of the free variable "x" to the left of the turnstile. This is accomplished by applying the deduction principle. A first application yields

$$(5.39)\qquad P, L(x) \vdash H(x) \rightarrow (x \text{ lies in } \pi).$$

Generalization is still not legitimate, and if incorrectly used would produce

> $\pi \perp$ b at A and x is a line yields, for everything, if it is perpendicular to b at A then it lies in π,

which is still absurd. Applying the deduction principle to (5.39) produces

$$(5.40)\qquad P \vdash L(x) \rightarrow [H(x) \rightarrow (x \text{ lies in } \pi)].$$

Since in (5.40), "x" is not free in "P", the generalization principle may be used to obtain

$$(5.41)\qquad P \vdash (x)\{L(x) \rightarrow [H(x) \rightarrow (x \text{ lies in } \pi)]\}.$$

In translation (5.41) is read

> $\pi \perp$ b at A yields, for every x, whenever x is a line if $x \perp$ b at A then x lies in π.

Or more comfortably,

> $\pi \perp$ b at A yields, every line which is perpendicular to b at A lies in π.

One more application of the deduction principle to (5.41) yields the required theorem

$$(5.42)\qquad \vdash P \rightarrow (x)\{L(x) \rightarrow [H(x) \rightarrow (x \text{ lies in } \pi)]\}\dagger$$

The valid formula (2.55) used in the extended sense can be applied to the statement function in braces to reduce the number of conditional symbols in (5.42):

$$(5.43)\qquad \vdash P \rightarrow (x)\{L(x)H(x) \rightarrow (x \text{ lies in } \pi)\}$$

In translation, (5.43) is read

> If $\pi \perp$ b at A, then every line perpendicular to b at A lies in π.

With the inference rule IU and the principle PGU*, formal machinery exists for proving the validity of such syllogisms as the following:

† Absence of any symbols to the left of the turnstile indicates that there is a demonstration of the statement on the right of the turnstile from the axioms of solid geometry.

All integers are real numbers.
All real numbers are complex numbers.
∴ All integers are complex numbers.

Translating in an obvious way,

$$
\begin{aligned}
&(x)[I(x) \rightarrow R(x)] \\
&(x)[R(x) \rightarrow C(x)] \\
\therefore\ &(x)[I(x) \rightarrow C(x)].
\end{aligned} \tag{5.44}
$$

To establish the validity of (5.44), we have to exhibit a demonstration that the conclusion follows from the premises, that is, to establish

$$(x)[I(x) \rightarrow R(x)],\ (x)[R(x) \rightarrow C(x)] \vdash (x)[I(x) \rightarrow C(x)].$$

Demonstration:

1. $(x)[I(x) \rightarrow R(x)]$	1. Hyp
2. $I(y) \rightarrow R(y)$	2. IU
3. $(x)[R(x) \rightarrow C(x)]$	3. Hyp
4. $R(y) \rightarrow C(y)$	4. IU
5. $I(y) \rightarrow C(y)$	5. Extended sense of inference rule (2.75) applied to Steps 2 and 4.
6. $(x)[I(x) \rightarrow C(x)]$	6. PGU*

Note that the application of PGU* is legitimate here since there is no free "x" in the assumptions on which the argument is based. The argument

Every womrath outgrabes.
Nothing that outgrabes is a real creature.
∴ No womrath is a real creature.

with translation

$$
\begin{aligned}
&(x)[W(x) \rightarrow O(x)] \\
&(x)[O(x) \rightarrow \sim C(x)] \\
\therefore\ &(x)[W(x) \rightarrow \sim C(x)]
\end{aligned}
$$

is demonstrated valid as follows:

1. $(x)[W(x) \rightarrow O(x)]$	1. Hyp
2. $W(y) \rightarrow O(y)$	2. IU
3. $(x)[O(x) \rightarrow \sim C(x)]$	3. Hyp
4. $O(y) \rightarrow \sim C(y)$	4. IU
5. $W(y) \rightarrow \sim C(y)$	5. (2.75) applied to Steps 2 and 4
6. $(x)[W(x) \rightarrow \sim C(x)]$	6. PGU*

EXERCISES

Write a demonstration for each of the following arguments:

1. Every rational is a real number.
No real number is imaginary.
∴ No rational number is imaginary.

2. All human beings are living things.
No living thing is perfect.
Every one of us is a human being.
∴ None of us is perfect.

3. All irrational numbers are real.
All real numbers are complex numbers.
∴ All irrational numbers are complex numbers.

Use the symbolic translations of statement functions given below, and apply the generalization principle PGU* in each of the exercises.

$T(x)$: x is a triangle.
$Is(x)$: x is isosceles.
$S(x)$: x has two equal line segments.
$A(x)$: x has two equal angles.
$Eq(x)$: x has three equal line segments.

4. $T(x),\ Is(x),\ S(x) \vdash A(x)$
5. $T(x),\ A(x) \vdash S(x)Is(x)$
6. $T(x),\ Is(x),\ S(x) \vdash \sim Eq(x)$
7. $T(x),\ Eq(x) \vdash Is(x)$

5.11 INFERENCE OF A STATEMENT FUNCTION FROM AN EXISTENTIAL STATEMENT: IE

Proof of the validity of arguments containing particular (existential) statements requires further formal machinery. The argument

All trigonometric functions are periodic functions.
Some trigonometric functions are continuous.
∴ Some periodic functions are continuous.

is valid, and may be translated

$$\begin{array}{ll} & (x)[T(x) \rightarrow P(x)] \\ (5.45) & (\exists x)[T(x)C(x)] \\ & \therefore\ (\exists x)[P(x)C(x)]. \end{array}$$

If we attempt to establish validity as in the previous examples, we easily get the first three steps:

1. $(x)[T(x) \rightarrow P(x)]$	1. Hyp
2. $T(y) \rightarrow P(y)$	2. IU
3. $(\exists x)[T(x)C(x)]$	3. Hyp

We have no rule as yet to infer any statement function from Step 3, but might like to continue:

4. $T(y)C(y)$	4. ?
5. $T(y)$	5. Conj simp, (2.73)
6. $C(y)$	6. Similarly
7. $P(y)$	7. Mod pon [2,5]
8. $P(y)C(y)$	8. Conj inf (2.72)
9. $(\exists x)[P(x)C(x)]$	9. ?

Going from Step 3 to Step 4 requires some sort of inference scheme such as

$$\frac{(\exists x)F(x)}{F(y)}. \tag{5.46}$$

The inference seems reasonable enough, since we interpret "$(\exists x)F(x)$" to mean that there exists at least one true substitution instance of "$F(x)$". Clearly, however, there is a fundamental difference between the scheme (5.46) and IU. In inferring "$F(y)$" from "$(x)F(x)$", "y" is quite unrestricted, but "y" in the "$F(y)$" of (5.46) clearly requires more delicate consideration. "$(\exists x)F(x)$" is interpreted as meaning that "$F(x)$" has at least one true substitution instance. It does not follow from this that any particular substitution we choose will be a true one. Indeed, it does not follow that we can even exhibit a true substitution instance of a given true existential statement. Consider

$$F(x)\colon x \text{ is a prime number greater than } 2^{1279} - 1.$$

It has been known since Euclid that there is no largest prime number. Hence, it is known that

$$(\exists x)F(x)$$

is a true statement. At this writing, no one has exhibited a prime number larger than $2^{1279} - 1$.

Since all the inference rules are devices needed to construct demonstrations, it is reasonable to handle the difficulties surrounding the inference scheme (5.46), inference of a statement function from an existential statement, by attaching to its definition restrictions on its use in demonstrations. We now formally state the rule with its restrictions:

(5.47) IE: For any statement function $F(x)$,

$$\frac{(\exists x)F(x)}{F(y)}$$

where "y" is not a free variable in any step of the demonstration prior to this inference, and "y" is not a free variable in any assumption formula (to the left of $\vdash$).

Looking back at the demonstration of (5.45), we see that the inference of Step 4 from Step 3 violates the restrictions on the use of IE since "y" is free in Step 2. However, the restrictions on IE in Step 3 do not hold for IU; so we can salvage the demonstration by reshuffling the first few steps:

1. $(\exists x)[T(x)C(x)]$	1. Hyp
2. $T(y)C(y)$	2. IE
3. $(x)[T(x) \to P(x)]$	3. Hyp
4. $T(y) \to P(y)$	4. IU

and so on, as before.

With the addition of IE to the machinery of demonstration it is necessary to add a restriction to PGU*, namely, that one must not try to generalize on a variable introduced by applying IE. Without such a restriction, we could prove the absurd statement

(5.48) $$\vdash (\exists x)P(x) \to (x)P(x)$$

as follows:

1. $(\exists x)P(x)$	1. Hyp
2. $P(y)$	2. IE
3. $(x)P(x)$	3. PGU*
4. $(\exists x)P(x) \to (x)P(x)$	4. Ded prin

This absurd result arises from a misuse of PGU* to obtain Step 3 from Step 2. At Step 2 in this faulty demonstration, it has been correctly established that

$$(\exists x)P(x) \vdash P(y).$$

Now "y" is not a free variable in the assumption "$(\exists x)P(x)$"; so PGU* as stated in (5.36) could be thought to apply. But an application of PGU* at this point leads to the absurd result that if at least one thing has property P, all things do.

To avoid this fallacy, we add to PGU* a restriction. A restatement of PGU* with the added restriction follows:

(5.49) PGU*: For any statement function $F(x)$, if $A_1, A_2, \ldots, A_m \vdash F(x)$, then $A_1, A_2, \ldots, A_m \vdash (x)F(x)$, where each of the A's may be a statement or statement function, but x is not a free variable in any of the A's, nor is x a variable *freed* by use of IE.

The variable "y" in the foregoing fallacious demonstration is such a *freed* variable since its introduction as a free variable was due to the use of IE.

The added restriction in (5.49) proves adequate as long as statement functions of one variable are involved but requires further modification for statement functions of more than one variable. We shall delay stating PGU* in its final form until functions of more than one variable have been treated.

5.12 INTRODUCTION OF THE EXISTENTIAL QUANTIFIER

To complete the formal machinery necessary to demonstrate (5.45), we need a procedure for getting from Step 8 to Step 9. The principle may be expressed in a form somewhat similar to PGU*, and indeed can be established from the generalization principle. We refer to it in the future by the letters PGE, which we can think of as an abbreviation for *Principle of Generalization to an Existential Statement.*

(5.50) PGE: For any function $F(x)$, if
$A_1, \ldots, A_n \vdash F(x)$, then
$A_1, \ldots, A_n \vdash (\exists x)F(x)$.

This can be established by starting with:

1. $(x){\sim}F(x)$	1. Assumption
2. ${\sim}F(x)$	2. IU

So we have proved

$$(x){\sim}F(x) \vdash {\sim}F(x),$$

and by the deduction principle,

$$\vdash (x){\sim}F(x) \rightarrow {\sim}F(x),$$

and by contrapositive inference,

$$\vdash F(x) \rightarrow {\sim}(x){\sim}F(x).$$

But by (5.24*d*)

$${\sim}(x){\sim}F(x) \leftrightarrow (\exists x)F(x),$$

so by substitution,

(5.51) $$\vdash F(x) \rightarrow (\exists x)F(x).$$

Thus, if we have

$$A_1, \ldots, A_n \vdash F(x),$$

then, using (5.51) and modus ponens, we obtain

(5.52) $$A_1, \ldots, A_n \vdash (\exists x)F(x).$$

5.13 USE OF THE QUANTIFICATION INFERENCE RULES

The inference schemes and principles IU, PGU*, IE, PGE are devices used in demonstrations to take off and put on quantifiers. They allow us to arrange proofs so that the majority of steps deal with unquantified statement functions. Most of the theorems we want to prove in mathematics are statements, and in their symbolic form, all variables are bound. The axioms and previously proved theorems on which the proofs are based are also fully quantified, i.e., all variables are bound. With the new inference rules, we can "unquantify" any needed axioms or theorems, and when the demonstration attains the right statement function, we may quantify to obtain the desired theorem.

As we have seen, removing a universal quantifier (IU) or introducing an existential quantifier (PGE) are both unrestricted. Whereas, introducing a universal quantifier (PGU*) or removing an existential quantifier (IE) are hedged about with restrictions.

In the following sample proofs we illustrate use of these principles and rules as well as their misuse.

Example 1. Prove validity of the quantified statement:[1]

$$\vdash (\exists x)[P(x)Q(x)] \to [(\exists x)P(x)(\exists x)Q(x)]$$

1. $(\exists x)[P(x)Q(x)]$	1. Assumption
2. $P(y)Q(y)$	2. IE, Step 1
3. $P(y)$	3. Conj simp, Step 2
4. $Q(y)$	4. Conj simp, Step 2
5. $(\exists x)P(x)$	5. PGE, Step 3
6. $(\exists x)Q(x)$	6. PGE, Step 4
7. $(\exists x)P(x)(\exists x)Q(x)$	7. Conj inf, Steps 5 and 6

At this stage we have demonstrated

$$(\exists x)[P(x)Q(x)] \vdash (\exists x)P(x)(\exists x)Q(x).$$

So by the deduction principle

$$\vdash (\exists x)[P(x)Q(x)] \to [(\exists x)P(x)(\exists x)Q(x)].$$

We check on the use of IE by observing that "y" in Step 2 is not free in any earlier step.

Example 2. In contrast, we illustrate a fallacious demonstration of the invalid formula:

$$(5.53) \qquad (\exists x)P(x)(\exists x)Q(x) \to (\exists x)[P(x)Q(x)]$$

[1] Note that the right side could as well be written:

$$(\exists x)P(x)(\exists y)Q(y).$$

As an example of a choice for "$P(x)$", "$Q(x)$" for which (5.53) may be false, take for translations

$P(x)$: x is a mid-point of segment AB.
$Q(x)$: x is a mid-point of segment CD.

Then (5.53) asserts: If AB and CD each have mid-points, then they both have a common mid-point. By an axiom of geometry, any segment has a mid-point, but it is easy to choose segments AB and CD whose mid-points are distinct. For this choice of segments the conditional (5.53) has a true antecedent and false consequent and is thus false.

Proof of (5.53) [FALLACIOUS]:

1. $(\exists x)P(x)(\exists x)Q(x)$	1. Assumption
2. $(\exists x)P(x)$	2. Conj simp, Step 1
3. $(\exists x)Q(x)$	3. Conj simp, Step 1
4. $P(x)$	4. IE, Step 2
5. $Q(x)$	5. IE, Step 3
6. $P(x)Q(x)$	6. Conj inf, Steps 4 and 5
7. $(\exists x)[P(x)Q(x)]$	7. PGE, Step 6

and (5.53) follows by the deduction principle.

The fallacy lies in the misuse of IE at Step 5. In Step 4, IE is used correctly, since "x" is not free in Steps 1, 2, and 3. [We could have written "$P(y)$" for Step 4 instead of "$P(x)$" but are not obliged to.] In Step 5, we may not use IE to obtain "$Q(x)$" since "x" is free in Step 4. Here we are obliged to use some variable other than "x". Then the demonstration might proceed:

5. $Q(y)$	5. IE, Step 3
6. $P(x)Q(y)$	6. Conj inf
7. $(\exists x)[P(x)Q(y)]$	7. PGE

In Step 7, "y" is still free. We might apply PGE again to bind "y", obtaining

8. $(\exists y)(\exists x)[P(x)Q(y)]$	8. PGE

and then use the deduction principle to obtain

$$(\exists x)P(x)(\exists x)Q(x) \rightarrow (\exists y)(\exists x)[P(x)Q(y)]. \tag{5.54}$$

But, of course, this is not (5.53).

Example 3. Prove: $\vdash (x)F(x) \rightarrow (\exists x)F(x)$

1. $(x)F(x)$	1. Assumption
2. $F(y)$	2. IU, Step 1
3. $(\exists x)F(x)$	3. PGE, Step 2
4. $(x)F(x) \rightarrow (\exists x)F(x)$	4. Ded prin

Example 4. Prove: $\vdash (\exists x)F(x) \rightarrow (x)F(x)$ [This is invalid.]

1. $(\exists x)F(x)$	1. Assumption
2. $F(y)$	2. IE
3. $(x)F(x)$	3. PGU*
4. $(\exists x)F(x) \rightarrow (x)F(x)$	4. Ded prin

Of course, here "y" in Step 2 was freed by IE; hence, we may not generalize on it in Step 3.

Example 5. Suppose that we have already proved that the square of an even integer is an even integer, and we wish to prove that if the square root of an odd integer is itself an integer, then it is odd.

As translations take

$I(x)$: x is an integer
$E(x)$: x is even (for "x is odd" take "$\sim E(x)$")

Then we take as a theorem already proved

$$T_1: (x)\{I(x) \rightarrow [E(x) \rightarrow I(x^2)E(x^2)]\}.$$

We want to prove

$$T_2: (x)\{I(x^2) \rightarrow [\sim E(x^2)I(x) \rightarrow \sim E(x)]\}.$$

The proof is indirect. We start with "$\sim T_2$" and deduce a contradiction. Hence start with

1. $\sim(x)\{I(x^2) \rightarrow [\sim E(x^2)I(x) \rightarrow \sim E(x)]\}$	1. Hyp
2. $(\exists x)\sim\{I(x^2) \rightarrow [\sim E(x^2)I(x) \rightarrow \sim E(x)]\}$	2. Substitution in Step 1 from (5.24*a*)
3. $\sim\{I(y^2) \rightarrow [\sim E(y^2)I(y) \rightarrow \sim E(y)]\}$	3. IE
4. $I(y^2)\sim[\sim E(y^2)I(y) \rightarrow \sim E(y)]$	4. Subst. in Step 3 from (2.41)
5. $I(y^2)\sim E(y^2)I(y)\sim\sim E(y)$	5. Subst. in Step 4 from (2.41)
6. $I(y)$	6. Conj simp, Step 5
7. $\sim\sim E(y)$	7. Conj simp, Step 5
8. $E(y)$	8. Subst. in Step 7 from (2.40)
9. $\sim E(y^2)$	9. Conj simp, Step 5

Now we make use of the theorem "T_1":

10. $I(y) \rightarrow [E(y) \rightarrow I(y^2)E(y^2)]$	10. IU from T_1
11. $E(y) \rightarrow I(y^2)E(y^2)$	11. Mod pon [6,10]
12. $I(y^2)E(y^2)$	12. Mod pon [8,11]

13. $E(y^2)$	13. Conj simp, Step 12
14. $E(y^2){\sim}E(y^2)$	14. Conj inf, Steps 9, 13

With the contradiction of Step 14, "$\sim T_2$" is proved false, and "T_2" is proved.

EXERCISES

With suitable symbolic translations of statement functions, write a demonstration for each of the following arguments:

1. All real numbers are rational or irrational.
Some real numbers have a repeating infinite decimal representation.
∴ Some rational or irrational numbers have a repeating infinite decimal representation.

2. All rectangles with four equal sides are squares.
Some rectangles have four equal sides.
∴ Some rectangles are squares.

3. All men who smoke cigarettes inhale.
Some men who smoke cigarettes develop cancer of the lungs.
∴ Some men who smoke cigarettes and inhale develop cancer of the lungs.

4. It is not the case that all angles can be trisected with straightedge and compass.
∴ There exists an angle that cannot be trisected with straightedge and compass.

5. Since there exists an angle that is trisectable by straightedge and compass, it is not the case that not all angles are not trisectable by straightedge and compass.

Write out a demonstration for each of the following:

6. $\vdash (\exists x)F(x) \leftrightarrow (\exists y)F(y)$

7. $\vdash (\exists x)P(x)(x)Q(x) \rightarrow (\exists x)[P(x)Q(x)]$

5.14 STATEMENT FUNCTIONS OF SEVERAL VARIABLES

So far, with minor exceptions, we have treated simple and compound statement functions involving only one variable, and have never applied more than one quantifier at a time to a statement function. Now taking as translations,

$R(x)$: x is a rational number,
$P(x)$: x is a prime number,

consider the statement

$$R(4) \vee P(7). \tag{5.55}$$

We view (5.55) as a substitution instance of the compound statement function of two variables

$$R(x) \vee P(y). \tag{5.56}$$

One cannot obtain (5.55) as a substitution instance of

$$R(x) \vee P(x), \tag{5.57}$$

since the conventions require that the same name replace "x" wherever it occurs in a statement function. From (5.57) we could obtain substitution instances:

$$R(4) \vee P(4), \qquad R(7) \vee P(7), \qquad R(a) \vee P(a), \qquad R(3) \vee P(3). \tag{5.58}$$

All the instances (5.58) are also substitution instances of (5.56), since, while we are permitted to replace "x" and "y" in (5.56) by different names, we are not required to do so.

In the same way, the set of substitution instances of

$$A(x) \rightarrow B(y) \tag{5.59}$$

contains all the substitution instances of "$A(x) \rightarrow B(x)$", along with other instances that cannot be obtained from "$A(x) \rightarrow B(x)$".

If we quantify (5.59) once by "(x)", we obtain the statement function of one free variable

$$(x)[A(x) \rightarrow B(y)], \tag{5.60}$$

and if this is quantified by "(y)", we obtain the statement

$$(y)\{(x)[A(x) \rightarrow B(y)]\}. \tag{5.61}$$

In (5.61) the scope of "(x)" is the statement function (5.59), and the scope of "(y)" is the statement function (5.60).

Normally we omit the braces in (5.61) and write it

$$(y)(x)[A(x) \rightarrow B(y)]$$

or even

$$(y,x)[A(x) \rightarrow B(y)].$$

Of course, if the scope of "y" is to include statement functions in addition to (5.60), the braces will be needed to indicate the scope. Compare

$$(y)(x)[A(x) \rightarrow B(y)] \vee C(y)$$

and

$$(y)\{(x)[A(x) \rightarrow B(y)] \vee C(y)\}.$$

In the first of these the third "y" is free since the scope of any quantifier is just that statement function immediately to its right, which is (5.60) in this case, so that "y" in "$C(y)$" is not bound. The second expression has all variables bound, and is a statement.

If we quantify (5.59) first by "$(\exists x)$" and then by "$(\exists y)$", we would again obtain a statement and would write it

$$(\exists y)(\exists x)[A(x) \rightarrow B(y)]$$

or sometimes

$$(\exists y,x)[A(x) \rightarrow B(y)].$$

Let us take "$F(x,y)$" as a convenient abbreviation for the compound statement function (5.59), then we may list all the ways in which (5.59) may be quantified so as to bind both variables, as follows:

$$(x)(y)F(x,y)$$
$$(y)(x)F(x,y)$$
$$(x)(\exists y)F(x,y)$$
$$(\exists y)(x)F(x,y)$$
$$(\exists x)(y)F(x,y)$$
$$(y)(\exists x)F(x,y)$$
$$(\exists x)(\exists y)F(x,y)$$
$$(\exists y)(\exists x)F(x,y)$$

While these are all formally distinct, some of them are logically equivalent as we shall see farther on.

Just as there are statements that cannot be expressed as compounds of simple statements, there are statement functions of two variables that cannot be expressed as compounds of statement functions of one variable. The statement

(5.62) 8 and 15 are integers,

can be rephrased

8 is an integer and 15 is an integer.

Then with the translation

$I(x)$: x is an integer,

an adequate translation of (5.62) is

(5.63) $$I(8)I(15).$$

We can regard (5.63) as a substitution instance of the statement function

"$I(x)I(y)$". In contrast, the statement

(5.64) 8 and 15 are relatively prime†

cannot be similarly treated. It makes no sense to write

8 is relatively prime and 15 is relatively prime.

Indeed "8 is relatively prime" is not a sentence. The difference between (5.62) and (5.64) is that (5.62) asserts simply that 8 and 15 have a common property, namely, each is in the class of integers, whereas (5.64) states that 8 and 15 are *related* in a certain way. "Relative primeness" is not a property that a single integer can have, but is a relationship that may exist between a pair of integers. We can take as a statement function of two variables

(5.65) $F(x,y)$: x and y are relatively prime.

Then, translate (5.64)

$$F(8,15).$$

However, "$F(x,y)$" cannot be expressed as a compound of statement functions of one variable.

The following statement is a nonmathematical example of a relation:

(5.66) John and Joan are engaged.

To say

John is engaged and Joan is engaged

makes sense in terms of current usage of "engaged", but the sense is not that of (5.66), which clearly means that John and Joan are engaged to each other.

Relationships between two or more unknowns abound in mathematics. Among statement functions of algebra expressing relations between numbers are:

a. $x < y$
b. $x = y$
c. $x + y = 10$
d. $x + y = z$
e. $0 < x < y \leq 1$
f. x and y are integers and x divides y
g. $x + y$ is a factor of $x^3 + y^3$
h. $|x + y| \leq |x| + |y|$
i. $(x + y) + z = 1$
j. $x^2 + y^2 = z^2$

In Euclidean geometry it is difficult to find any statements which do not express relationships. Even in our miniature geometry, Sec. 4.2,

† Two integers are relatively prime if and only if they have no common divisor other than 1 and −1.

which has only one undefined relational word "on", only one of the axioms and definitions does not involve a relation. It is

A_1: There exists at least one line.

If we use "$L(x)$" as a translation of "x is a line", then we can translate

$$A_1\text{: } (\exists x)L(x).$$

Axiom A_0 of the geometry involves the relation "on". With translations:

$P(x)$: x is a point
$L(x)$: x is a line
$W(x,y)$: x is on y

we may translate

A_0: Point P is on line q if and only if line q is on point P,

which is clearly meant to be a general statement about all points and lines of the geometry, by the quantified statement

$$(x,y)\{P(x)L(y) \rightarrow [W(x,y) \leftrightarrow W(y,x)]\}.$$

Among statement functions of relationship in Euclidean geometry are:

a. Line segment AB has a mid-point M.
b. Point P is on line l.
c. l $\|$ m (l and m are lines)
d. $\angle A < \angle B$
e. A is equidistant from B and C.
f. $\triangle ABC \cong \triangle A'B'C'$
g. Side b is adjacent to $\angle A$.
h. Point A lies between points B and C on line a.
i. Lines a and b are skew.
j. In a triangle with sides a, b, c, $a + b > c$.

EXERCISES

In each of the following exercises give (*a*) a somewhat literal translation of the expression and (*b*) a restatement in terms of conversational English. Use the translations that follow:

$P(x)$: x is a point.
$L(x)$: x is a straight line.
$W(x,y)$: y is on x.
$B(x,y,z)$: y is between x and z.
$M(xy,z)$: z is the mid-point of segment xy.

1. $P(\text{p})L(\text{l})W(\text{l,p})$
2. $P(\text{A})P(\text{B})P(\text{C})L(\text{a})W(\text{a,A})W(\text{a,B})W(\text{a,C})B(\text{B,A,C})$
3. $(x,y,z)\{(\exists w)[P(x)P(y)P(z)L(w)W(w,x)W(w,y)W(w,z)$
$$B(x,y,z)] \rightarrow B(z,y,x)\}$$
4. $(x,z)(\exists w)\{P(x)P(z)L(w)W(w,x)W(w,z) \rightarrow$
$$(\exists y)[P(y)W(w,y)B(x,y,z)](\exists v)[P(v)W(w,v)B(x,z,v)]\}$$

Using suitable translations, translate each of the following statements:

5. Point x is on line l.

6. Point z is on segment AB.

7. "If A, B, C are points of a straight line and B lies between A and C, then B lies also between C and A."†

8. "If A and C are two points of a straight line, then there exists at least one point B lying between A and C and at least one point D so situated that C lies between A and D."‡

Use the indicated translations to translate each of the following statements:

$N(x)$: x is a number	$Re(x)$: x is real
$I(x)$: x is integral	$C(x)$: x is complex
$E(x)$: x is even	$Pr(x)$: x is prime
$O(x)$: x is odd	$Rel(x,y)$: x is relatively prime to y
$Ra(x)$: x is rational	

9. All numbers are complex numbers.

10. Some real numbers are not rational.

11. There exists an integer that is an even prime number.

12. Two even numbers cannot be relatively prime.

13. Some rational numbers are odd and some rational numbers are even.

14. All rational numbers are real numbers.

15. Not all real numbers are rational.

16. All prime numbers are odd except 2.

5.15 EXTENSION OF QUANTIFICATION INFERENCE RULES

The inference rules of the statement calculus were extended to apply to statement functions of one variable in Sec. 5.8. In the same way, they may be further extended to apply to statement functions of several variables. With respect to the four quantification inference rules (Secs. 5.7, 5.10, 5.11, and 5.12) the extension must be made with care. In this

† David Hilbert, *The Foundations of Geometry*, p. 6, La Salle, Ill.: The Open Court Publishing Company, 1902.

‡ *Ibid.*

final extension, IU and PGE are again virtually unrestricted in application. The only new restrictions required are designed to avoid what may be called *confusion of bound variables.* Consider the statement

(5.67) $$(x)(\exists y)F(x,y).$$

Using IU, the following statement functions may be inferred from (5.67):

(5.68) $$(\exists y)F(x,y),\ (\exists y)F(z,y),\ (\exists y)F(t,y),$$

but not the statement

(5.69) $$(\exists y)F(y,y).$$

Suppose, for example, that the universe of individuals is restricted to the natural numbers, and take as translation

(5.70) $$F(x,y)\colon x < y,$$

then a translation of (5.67) is

$$(x)(\exists y)F(x,y)\colon \text{For every } x \text{ there is a } y \text{ such that } x < y,$$

which is true. Now a translation of (5.69) is

$$(\exists y)F(y,y)\colon \text{There exists a number } y \text{ such that } y < y,$$

which is false. The trouble here is that in "$(x)(\exists y)F(x,y)$" the first variable in "$F(x,y)$" is not bound by "$(\exists y)$", but in "$(\exists y)F(y,y)$" the first variable is bound by "$(\exists y)$". In the step from (5.67) to any of the functions (5.68), IU is used to free "x". In going from (5.67) to (5.69), "x" is not freed but becomes bound by a new quantifier. Such a step results in confusion of bound variables and is excluded under IU.

The same sort of confusion can arise if PGE is not restricted. From

(5.71) $$(x)F(x,y)$$

one may infer

$$(\exists z)(x)F(x,z)$$

by PGE, but not

(5.72) $$(\exists x)(x)F(x,x).$$

Similarly, from

(5.73) $$(\exists x)F(x,y)$$

one may infer

$$(\exists y)(\exists x)F(x,y)$$

by PGE, but not

(5.74) $$(\exists x)(\exists x)F(x,x).$$

In (5.71), the second variable of "$F(x,y)$" is not bound by the universal quantifier, but becomes bound by it in (5.72). The step from (5.73) to (5.74) is a second illustration of the same point.

To avoid this kind of confusion, one must be careful to ensure that no variable that is free of the quantifiers in a given expression becomes bound by one of these quantifiers after application of IU or PGE. We shall refer to this type of fallacy in the future as confusion of bound variables. Except for a statement about confusion of bound variables, the final forms for IU, PGE, IE are as already stated.

In its final form, PGU must be so restricted that confusion of bound variables is avoided in applications. One further restriction is necessary. The following fallacious demonstration will illustrate the necessity for restriction of PGU* when it is applied to statement functions of more than one variable. Consider the invalid formula

(5.75) $$(x)(\exists y)F(x,y) \to (\exists y)(x)F(x,y).$$

Let us first interpret this using the class of integers as the universe of individuals, and taking as translation

$$F(x,y): x < y.$$

Then (5.75) becomes on translation

> If for every integer there exists one which is greater,
> then there is an integer that is greater than all other integers.

The antecedent of this conditional is true, since if b is an arbitrary integer, then $b + 1$ is a greater integer. The consequent is false. One recognizes it as another way of saying that there is a greatest integer. In this interpretation, (5.75) is false. We can, however, apparently demonstrate (5.75) by applying PGU*:

1. $(x)(\exists y)F(x,y)$	1. Assumption
2. $(\exists y)F(z,y)$	2. IU
3. $F(z,w)$	3. IE

Now PGU*, as restricted by (5.49), will not allow generalization on "w" in Step 3 since "w" was freed by IE, but "z" was freed by IU, and so apparently:

4. $(x)F(x,w)$	4. PGU*
5. $(\exists y)(x)F(x,y)$	5. PGE
6. $(x)(\exists y)F(x,y) \to (\exists y)(x)F(x,y)$	6. Ded prin

The trouble occurs at Step 4. The "z" in Step 2 is completely unrestricted. Indeed, under the interpretation we have taken for "F", any substitution instance of Step 2 is a true statement. In Step 3, the "w" introduced by IE is restricted, as we have seen in other illustrations.

Under the present interpretation, "w" must denote an integer greater than z, but then of course "z" must denote an integer less than "w" and so "z" is also restricted. We must not try to generalize on "z" from any formula in which "w" is also present. If it were possible to continue the argument somehow so as to obtain a formula in which "z" occurs but "w" does not, then it would be correct to generalize on "z". To put it another way, "z" was not introduced by IE, but "w" was introduced by IE, and "z" was free in the formula inferred by IE. Hence, one may not generalize on "w" at any stage, but may generalize on "z" if "w" does not occur in the formula to which PGU* is to apply.

The restriction on PGU* necessary to avoid the type of fallacy occurring at Step 4 of the foregoing illustration can be stated

> Do not use PGU* to generalize on a variable that is free in any formula resulting from application of IE while a variable introduced by IE is present.

5.16 FINAL FORMS OF THE QUANTIFICATIONAL INFERENCE RULES

In the final statements of the four quantificational inference rules it should be understood that "$F(x)$" may stand for a statement function having not only free occurrences of "x", but of other variables as well. For example, "$F(x)$" might stand for

$$(z)[A(x,z) \rightarrow B(z,y)]. \tag{5.76}$$

When quantificational inferences involve formulas as complicated as (5.76), there is always the possibility of confusing bound variables, as discussed in Sec. 5.15. Hence, it is to be understood that such confusion is not allowed in the application of any of the following four rules.

IU: *Inference from Statement Function Having a Universal Quantifier.*
For any statement function "$F(x)$",

$$\frac{(x)F(x)}{F(y)}.$$

IE: *Inference from Statement Function Having an Existential Quantifier.*
For any statement function "$F(x)$",

$$\frac{(\exists x)F(x)}{F(y)}$$

where "y" is not a free variable in any step of the demonstration prior to this inference, and is not a free variable in any assumption formula (to the left of $\vdash$).

NOTE: IU includes the inference of "$F(s)$" from "$(x)F(x)$" where "s" is an individual constant, i.e., the name of some unique object. No such inference is permitted by IE; while it may be possible to obtain "$F(s)$" in a demonstration, it may not be inferred directly from "$(\exists x)F(x)$".

PGU: *Principle of Generalization to Universal Quantification.* For any statement function "$F(x)$", if

$$A_1, A_2, \ldots, A_n \vdash F(y),$$

then

$$A_1, A_2, \ldots, A_n \vdash (x)F(x),$$

where "y" is not a free variable in any of the "A's", and if "y" is free in any prior step of the demonstration resulting from use of IE, then no variable introduced by that use of IE occurs free in "$F(y)$".

PGE: *Principle of Generalization to Existential Quantification.* For any statement function "$F(x)$", if

$$A_1, A_2, \ldots, A_n \vdash F(y),$$

then

$$A_1, A_2, \ldots, A_n \vdash (\exists x)F(x).$$

NOTE ON PGU: The restriction on PGU is not hard to observe in practice. Suppose one reaches the formula "$F(x,y,z)$" in a demonstration, and wishes to generalize on "x". One then checks back through all the steps of the demonstration involving IE. If none of "x", "y", "z" was introduced by IE, PGU applies. If, say, "y" was introduced by IE at some step, then one must check to see that "x" did not occur free in that step before generalizing on "x".

5.17 USE OF EQUALITY

Equality was discussed briefly at the end of Sec. 1.3. There we stated how we intended to interpret such statements as

7 = VII
7/9 = 14/18
Mark Twain = Samuel Clemens.

The interpretation may be expressed: If "a" and "b" are names, then "$a = b$" means that "a" and "b" are names of the same object. In what follows, we shall refer to this as the *standard interpretation.* That other interpretations are common in elementary mathematics was indicated in Exercises 13 and 14 at the end of Sec. 1.4.

Elementary geometry provides a variety of examples of nonstandard interpretations of " = ". For instance,

(5.77) A triangle is isosceles if it has two equal sides.

Since a side of a triangle is a line segment, the standard interpretation of "equal" does not apply to (5.77). In general, when one writes

(5.78) $$AB = CD$$

in a context of plane geometry, "AB" and "CD" denote line segments, and (5.78) is understood to be shorthand for

$$\text{length } AB = \text{length } CD,$$

for which the standard interpretation is the correct one. Similarly,

(5.79) $$\triangle ABC = \triangle EFG$$

is shorthand for

$$\text{area } \triangle ABC = \text{area } \triangle EFG,$$

for which the standard interpretation is again correct. In the case of angles,

(5.80) $$\angle ABC = \angle EFG$$

means that the two angles have the same size, and does not ordinarily mean that "∠ABC" and "∠EFG" are names of the same angle.

In expressing the relation between angles and their intercepted arcs, it is common practice to use some special sign to indicate equivalence. For instance,

(5.81) $$\angle AOC \overset{\circ}{=} \overset{\frown}{AC},$$

which may be read, "∠AOC is equal in degrees to arc AC".

The statements (5.78) to (5.81) have the common characteristic that each makes an assertion about an object without naming it. Of course, such a thing is possible only because by agreements and by experience with such statements, we are able to determine the object whose name is tacit. It is not much more complicated to write for (5.78) to (5.81):

(5.78*) $$m(AB) = m(CD)$$
(5.79*) $$m(\triangle ABC) = m(\triangle EFG)$$
(5.80*) $$m(\angle ABC) = m(\angle EFG)$$
(5.81*) $$m(\angle AOC) = m(\overset{\frown}{AC}),$$

where "m(______)" is interpreted to mean "measure of ______". This device permits the standard interpretation for " = " in all these cases. Other notational devices may be used to distinguish between the name of an object and the name of the measure of an object. One might make the agreement that if "AB" names a line segment, then "$\overline{AB}$" names its length, and make similar agreements for angles, areas, arcs, etc.

Two further uses of " = " in plane geometry are worth noting. "Equal circles" are generally understood to be circles of equal radius. "Equal triangles" are taken to be congruent triangles by a few authors. Such a use of "equal triangles" is an unnecessary addition to the ambiguity of " = ", and seems to us quite indefensible.

Statements about geometric ratios use " = " in the standard sense. The statement

$$AB/A'B' = CD/C'D'$$

is interpreted to mean that "AB/A'B'" and "CD/C'D'" are names of the same ratio. The standard interpretation is also used in statements of equality involving numbers.

Where " = " is interpreted as meaning "equal in measure" in statements about line segments, angles, arcs, etc., it is probably unwise to use " = " in the standard sense in statements about lines and points and planes, and other unmeasurable geometric objects. It is not common, in plane geometry textbooks, to write "A=B" when A and B are points, or to write "a = b" when a and b are lines. Instead, the constructions "A coincides with B" and "a coincides with b" are used, or even "A falls on B" and "a falls on b". As an illustration, consider a very informal proof of the plane geometry theorem

> If a line is perpendicular to one of two parallels, then it is perpendicular to the other also.

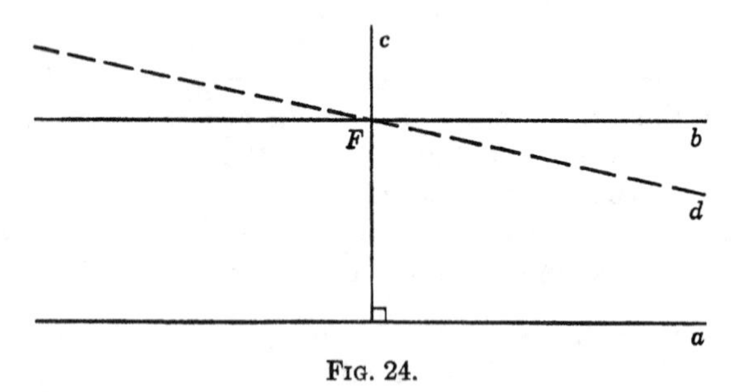

Fig. 24.

One starts (Fig. 24) with a $\parallel$ b and c $\perp$ a, and shows that since c meets a it must also meet b in some point, call it "F". Then there exists a line, call it "d", such that d $\perp$ c at F. By theorem, two lines perpendicular to a third are parallel; so d $\parallel$ a. Now, d and b both through F and both parallel to a implies d coincides with b, by the parallel postulate. Hence d coincides with b, and b ⊥ c.

The strategy of proof in the foregoing example may be summarized:

a. It is required to prove that b ⊥ c at F.
b. One proves the existence of a line d that is perpendicular to c at F.
c. One proves that d coincides with b.
d. One concludes that b ⊥ c at F.

A similar proof strategy is employed to prove

> If a line cuts two sides of a triangle in the same ratio, then the line is parallel to the third side.

We shall give a brief outline of a slight variant of the theorem. In the discussion, we shall have occasion to distinguish between a line segment AF and its length m(AF).

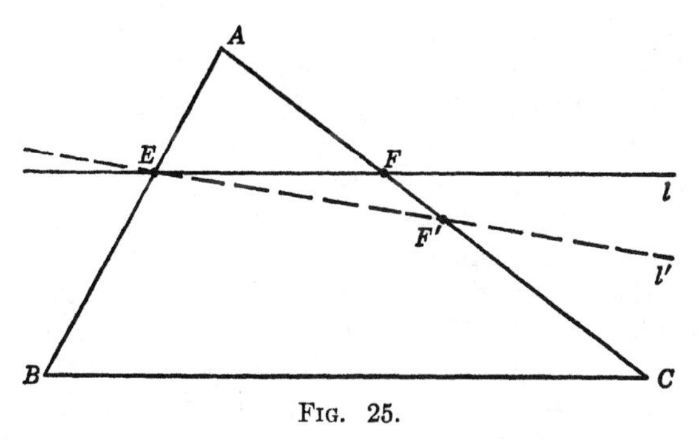

FIG. 25.

We start with △ABC and line l cutting AB at E and AC at F so that AE/AB = AF/AC (Fig. 25). We want to prove that l is parallel to the line through B and C. By the parallel postulate, there exists a line l′ through E that *is* parallel to line BC. Then it can be shown that l′ meets AC in some point, call it "F′". By theorem, AE/AB = AF′/AC. By a theorem on proportions, it follows that m(AF) = m(AF′), and by an axiom on line segments it follows that F coincides with F′. Hence by axiom, l coincides with l′, and thus l is parallel to the line through BC.

The proof strategy illustrated in the foregoing examples is common throughout mathematics. It is a goal of this section to develop the formal machinery to handle it. Before proceeding with this development, we shall discuss the use of *equality* in trigonometry and in algebra.

Much of the language of plane geometry is part of the language of trigonometry, as might be expected. Equality of line segments, arcs, angles, etc., is interpreted in the same way in both subjects. There are a variety of possible interpretations of statements of equality involving the trigonometric functions. Consider

$$\tan \alpha = \frac{a}{b}.$$

In this statement, "a" and "b" may denote either line segments, or lengths of line segments, or numbers having no connection at all with line segments. "α" may denote an angle, or the measure of an angle, or a number not connected in any way with an angle. (For the moment, we ignore the case where "α", "a", and "b" are placeholders.) In all these cases, "$\tan \alpha$" and "a/b" are names of numbers, and the standard interpretation of "=" is the correct one.

In a statement such as

$$c \sin \alpha = b,$$

it is hard to find a sensible interpretation if "b" and "c" denote line segments. If c is a line segment, what does "$c \sin \alpha$" denote? If "$c \sin \alpha$" is thought to be a line segment, what is its relation to c—or to b? It seems clear that the ambiguities of interpretation here are due to the notation "b" and "c" and not to "=".

The simplest way out is always to interpret "=" in the standard way for such equations, and to understand that "b" and "c" denote numbers. Depending on the context, the numbers can be lengths of line segments, or measures of other objects. If this way out is chosen, then a statement such as "A lies opposite a in $\triangle$ABC" is ambiguous in the use of "a".

It is quite possible to avoid such ambiguities in trigonometry by means of simple notational agreements. Using the following notational agreements, it should always be clear what the objects of discourse are. The suggested system of notation should not prove unwieldy, and has the advantage that "=" may be interpreted in the standard sense.

a. Small Latin letters always denote numbers, which in turn may be lengths of line segments.
b. Capital Latin letters always denote points.
c. "$\angle$ABC" always denotes an angle, and may be shortened to "$\angle$B" if no ambiguity results.
d. "AB" always denotes a line segment.
e. Small Greek letters always denote numbers, which in turn may be measures of angles.

Where it is desirable to indicate by the notation a correspondence between an angle and its measure, or a segment and its length, the following agreements might be adopted.

Geometric Object	*Corresponding Measure*
$\angle$ABC	β
$\angle$B	β
$\angle\beta$	β
AB	m(AB)
seg a	a

The suggested agreements are not ideal, but are put forward as an example of how the necessary distinctions can be made with moderate changes in current notation. A teacher wishing to observe the distinctions, but committed to a textbook that does not, will do well to work out his own notation with some regard to the notation of the textbook.

What we have called the standard interpretation of "=" is almost always correct in algebra. Of course, where the language of geometry or trigonometry occurs in algebra, ambiguities of interpretation will also occur. Since numbers are objects of discourse in elementary algebra, it must seem natural to a student to interpret "$a = b$" as meaning that a and b are the same number. The language used in geometry and, to some extent, in arithmetic has led him to think of "=" in terms of equality in measure, and measures are numbers. Yet the very triviality of the statement "$a = b$", where "a" and "b" each denote a number, can be confusing. In the language of geometry, the statement "AB = CD" really seems to have some meat in it, perhaps because "AB" and "CD" may denote distinct line segments, and the statement means that these segments have something about them that is the same. But in the case of "$a = b$", only a single number is mentioned, and what is asserted about it is trivial. The statement does have practical value, however, because from it one knows that "a" and "b" are different names for the same number. For example,

(5.82) The smallest proper prime factor of 4,294,967,297 = 641.

The left side of the equality (5.82) is another name for 641, yet a very fine mathematician did not know this, and indeed believed that "the smallest proper prime factor of 4,294,967,297" did not name any number (see Sec. 1.1.)[1]

Particularly in algebra there is need for a precise notion of equality, for a good part of the subject concerns statement functions involving the symbol "=". In the case of so-called *word problems*, the translation to algebraic language generally produces an algebraic equation. For example,

[1] There is a deeper problem involved here concerning the *denotation* and *connotation* of a name, which we do not pursue. The interested reader is referred to an excellent discussion of the matter in Alonzo Church, *Introduction to Mathematical Logic*, Princeton, N.J.: Princeton University Press, 1956.

PROBLEM: A man has $1.80 in nickels, dimes, and quarters. He has half as many quarters as dimes, and three times as many nickels as quarters. How many quarters has he?

The problem is easily translated into the algebraic formulation:

(5.83) Find all values of the variable that yield true substitution instances of the statement function:

$$(x \text{ is a positive integer})(25x + 20x + 15x = 180)$$

Observe that one cannot interpret "$25x + 20x + 15x$" as another name for 180, but that "$25x + 20x + 15x$" becomes the name of some number whenever "x" is replaced by a numeral, i.e., for every numerical substitution instance. Thus, every substitution instance of (5.83) is a statement in which equality is to be interpreted in the standard sense. That is, if "1/2" is substituted for "x" the result is a conjunction of two false statements, and is false. If "2" is substituted for "x", the result is a conjunction of a true statement and a false statement, and is false. It turns out that substitution of "3" for "x" produces the only true substitution instance; hence the answer to the problem is: The man has three quarters. If the algebraic reformulation of the problem is accepted as correct, then the answer is known to be correct because

$$25 \cdot 3 + 20 \cdot 3 + 15 \cdot 3 = 180$$

is a true statement. For the moment, we defer consideration of the uniqueness of this answer.

Another fundamental notion that frequently suffers from poor exposition is *substitution.* Not much can be done in algebra without an adequate substitution rule. A simple example is the solution of the simultaneous equations:

$$x + y = 10$$
$$x - y = 4$$

Using some axioms of algebra, the second equation is found to be equivalent to "$x = y + 4$", then the natural procedure is to write

$$(y + 4) + y = 10,$$

which represents a substitution into the first equation.

In student slang the substitution rule is sometimes stated

Equals may be substituted for equals,

which leaves a good deal tacit. The statement seems permissive, but it is not clear what is permitted, or what the results will be, or in what such substitutions are to be made. The students' slang is clearly inadequate, but their elders do not always improve on it. Efforts at stating a substitution rule often get involved with such words as "quantities" and "magnitudes"; such phrases as the following appear:

(5.84) . . . one quantity is replaced by an equal quantity in a statement . . .

or

(5.85) If two quantities are equal . . . .

The trouble with (5.84) is that quantities do not occur in statements (they are mentioned in statements) and hence cannot be replaced by equal quantities. The notion of "equal quantities" in (5.85) is confusing. A quantity is just itself, and it may be named by two distinct names, but the notion of "two quantities" requires that the quantities be distinguishable; hence, they cannot be equal in the standard interpretation.

The meaning of substitution is simple enough; a statement is unchanged in meaning if new names for objects are used in it, as long as the new names are synonyms for the names they replace. Observe that it is the names used that are replaced, and not the objects mentioned.

That the rule of substitution needs careful statement is illustrated by the following:

> Let ABC be isosceles with AC = BC. ABC has an altitude from A, call it "AD". Now AD $\perp$ BC; so by substitution, AD $\perp$ AC.

The argument is fallacious and the result absurd. It is left to the reader to state precisely what is wrong with this argument.

The foregoing discussion has been concerned with the meaning to be assigned to " = ". If " = " is to become a relational symbol of the logic, its manner of employment must be stated in terms of form alone. In particular, a substitution rule must be stated in terms of form alone. We know from the previous discussion what the interpretation is to be, and use it as a guide to the formal treatment of the next section.

5.18 FORMAL NOTION OF EQUALITY

In earlier sections we had occasion to use illustrative material from algebra, geometry, and trigonometry involving the relation of equality. We did not usually make explicit mention of the properties of " = ", but relied on the reader's experience of mathematics to fill in any tacit steps. An exception was the fairly careful demonstration of (3.34), where it was necessary to state and employ the transitive property of equality.

If " = " is regarded as a symbol of algebra, then one would expect to find among the axioms of algebra some specific axioms concerning the symbol. Three of the axioms are

(5.86) Reflexivity: $(x)[x = x]$

(5.87) Symmetry: $(x,y)[(x = y) \rightarrow (y = x)]$

(5.88) Transitivity: $(x,y,z)[(x = y)(y = z) \rightarrow (x = z)]$

Axiomatic development of algebra is not customary in secondary school mathematics, and it is rare to find these properties of equality explicitly discussed. Reflexivity and symmetry are tacit properties of equality in most derivations. In much of elementary mathematics, "equality" is not so much the name of a mathematical relation as it is a word of the underlying language like "implies" or "only if" and other words of the logic. In our formal treatment, we shall add the symbol " = " to the other symbols of the predicate calculus, and assume suitable inference rules regarding its use. In fact, we regard "$x = y$" as a binary statement function, and interpret a substitution instance of it to be true just in case names of a single object are substituted for "x" and "y", respectively.

We take as an axiom of the logic

(5.89) Reflexivity: $(x)[x = x]$

The axiom (5.89) will often be referred to by the abbreviation "R" in what follows.

We take as an inference rule, *inference by substitution* which will be referred to by the abbreviation IS in what follows.

(5.90) IS: If "$F(x)$" is a statement function and $F^*(y)$ is a statement function obtained from "$F(x)$" by replacing one or more free occurrences of the variable "x" by "y", then

$$\frac{x = y, F(x)}{F^*(y)}.$$

A comment is in order on the notation used to express (5.90). In all previous sections, where "$F(x)$" and "$F(y)$" appear in the same discussion, it is understood that "$F(y)$" is obtained by replacing *all* the free occurrences of "x" in "$F(x)$" by "y". Since it is not always desirable to substitute at all the free occurrences of "x", we adopted the symbol "$F^*(y)$" to denote the result of substitution in this case.

As with the inference rules PGU, PGE, IU, IE, there may be other free variables in "$F(x)$" besides "x". If in addition "$F(x)$" contains bound variables, then care must be taken to avoid confusion of bound variables in applying IS. An example of such a confusion is the following fallacious proof from algebra. To carry out the argument, we shall need a result about real numbers, namely: There is no largest real number. This may be stated symbolically:

$$(z)(\exists y)[z < y], \tag{5.91}$$

where it is understood that the universe of individuals is restricted to the real numbers.

Fallacious demonstration:

1. $x = y$	1. Assumption
2. $(z)(\exists y)[z < y]$	2. (5.91)
3. $(\exists y)[x < y]$	3. IU, Step 2
4. $(\exists y)[y < y]$	4. IS (incorrect use)

The fallacy arises in the inference from Step 3 to Step 4. In Step 3, "x" is free, but since it is in the scope of "$\exists y$" one may not replace it by "y" since this changes the status of the first variable in "$x < y$" from free to bound.

The properties of reflexivity, symmetry, and transitivity were discussed in this section in an algebraic context. The axiom R and the inference rule IS were introduced formally, however, and are thus not limited to any such context. By means of R and IS, it is possible to give formal demonstrations of (5.87) and (5.88); hence, symmetry and transitivity are also free of any special context, algebraic or other. The demonstrations follow:

Symmetry: $(x,y)[(x = y) \rightarrow (y = x)]$

(5.92)

1. $u = v$	1. Assumption
2. $(x)(x = x)$	2. R
3. $u = u$	3. IU, Step 2
4. $v = u$	4. IS, Steps 1 and 3
5. $(u = v) \rightarrow (v = u)$	5. Ded prin, Steps 1 and 4
6. $(x,y)[(x = y) \rightarrow (y = x)]$	6. PGU (applied twice to Step 5)

Transitivity: $(x,y,z)[(x = y)(y = z) \rightarrow (x = z)]$

(5.93)

1. $(u = v)(v = w)$	1. Assumption
2. $u = v$	2. Conj simp, Step 1
3. $v = w$	3. Conj simp, Step 1
4. $u = w$	4. IS, Steps 2 and 3
5. $(u = v)(v = w) \rightarrow (u = w)$	5. Ded prin, Steps 1 and 4
6. $(x,y,z)[(x = y)(y = z) \rightarrow (x = z)]$	6. PGU (applied three times to Step 5)

The next logical theorem follows easily from IS. We state it without proof:

$$(x,y)[(x = y) \to (F(x) \to F^*(y))], \tag{5.94}$$

where "$F(x)$" and "$F^*(y)$" are as in (5.90).

A useful contrapositive form of (5.94) is easily proved:

$$(x,y)[F(x){\sim}F^*(y) \to (x \neq y)]. \tag{5.95}$$

The result (5.95) may be paraphrased: If a statement function can be found that is true for "x" and false for "y", then $x \neq y$. The procedure of this paraphrase was employed informally many times in the Miniature Geometry, Sec. 4.2, to prove that two lines were distinct, or that two points were distinct. A simple illustration from plane geometry follows.

Suppose A is a point not on line b, and c is a line through A. It is required to prove that b and c are distinct lines. The need to prove such a thing as this is usually overlooked in plane geometry, probably because it is so obvious from a figure that b and c are distinct. If a proof was thought necessary, it would probably be indirect, starting from the assumption that b and c coincide and leading to the contradiction that A is both on and off c. A formal direct proof is possible, and illustrates the use of (5.95).

Take as translation

$W(x,y)$: x is a point on line y.

Then the translation of the given conditions is

$${\sim}W(\mathrm{A},\mathrm{b})\,W(\mathrm{A},\mathrm{c}).$$

By IU, from (5.95),

$$W(\mathrm{A},\mathrm{c}){\sim}W(\mathrm{A},\mathrm{b}) \to (\mathrm{c} \neq \mathrm{b});$$

hence by commutativity and modus ponens,

$$\mathrm{c} \neq \mathrm{b}.$$

In long demonstrations such as will be found in Chap. VI, the opportunity to employ (5.95) occurs so often that many steps can be saved by adopting an inference rule corresponding to it. The inference rule will be referred to by the abbreviation ISC, *contrapositive form of* IS, in what follows.

(5.96) ISC: If "$F(x)$" is a statement function and "$F^*(y)$ is a statement function obtained from "$F(x)$" by replacing one or more free occurrences of the variable "x" by "y", then

$$\frac{F(x),\ {\sim}F^*(y)}{x \neq y}.$$

5.19 THE SCOPE OF AN ASSUMPTION IN A DEMONSTRATION

In illustrative formal demonstrations and informal proofs of earlier sections, very often the final step has been justified by the deduction principle. In general, the deduction principle has to be used to facilitate proofs of conditional statements. That is, instead of trying to demonstrate

$$\vdash P \rightarrow Q, \tag{5.97}$$

we first demonstrated

$$P \vdash Q; \tag{5.98}$$

then obtained (5.97) by the deduction principle. Several applications of the deduction principle may be needed for some proofs. For example, to prove

$$\vdash P \rightarrow (Q \rightarrow R), \tag{5.99}$$

we start by assuming "P" and "Q" and then demonstrate

$$P,\ Q \vdash R. \tag{5.100}$$

Then (5.99) is obtained by two successive applications of the deduction principle. At any step of the demonstration of (5.100), either "P" or "Q" may be the next step. In long proofs, however, it is frequently convenient to make additional assumptions. That is, it may be easier to prove

$$P,\ Q,\ A,\ B \vdash S \tag{5.101}$$

first. Then by two applications of the deduction principle obtain

$$P,\ Q \vdash A \rightarrow (B \rightarrow S). \tag{5.102}$$

Then if "$A \rightarrow (B \rightarrow S)$" is a translation for "$R$", (5.100) is established, and (5.99) follows as before. Unfortunately, when starting on a long demonstration, it is usually not clear what additional assumptions will turn out to be convenient. For example, suppose it is required to demonstrate

$$\vdash (\exists x)[F(x) \vee G(x)] \rightarrow (\exists x)F(x) \vee (\exists y)G(y). \tag{5.103}$$

The first step would be to demonstrate

$$(\exists x)[F(x) \vee G(x)] \vdash (\exists x)F(x) \vee (\exists y)G(y). \tag{5.104}$$

Proof:

1. $(\exists x)[F(x) \vee G(x)]$	1. Assumption
2. $F(z) \vee G(z)$	2. IE, Step 1

At this point, one looks ahead and sees that if one could prove

(5.105) $$F(z) \to (\exists x)F(x) \vee (\exists y)G(y),$$

and

(5.106) $$G(z) \to (\exists x)F(x) \vee (\exists y)G(y),$$

then inference by cases and modus ponens (using Step 2) will deliver the proof. The easiest way to establish (5.105) is to assume "$F(z)$" and deduce "$(\exists x)F(x) \vee (\exists y)G(y)$"; then get (5.105) by the deduction principle. Thus, to demonstrate (5.104), one first demonstrates

(5.107) $$(\exists x)[F(x) \vee G(x)],\ F(z) \vdash (\exists x)F(x) \vee (\exists y)G(y).$$

The same sort of remark applies to establishing (5.106), and suggests that we have to establish

(5.108) $$(\exists x)[F(x) \vee G(x)],\ G(z) \vdash (\exists x)F(x) \vee (\exists y)G(y)$$

before getting (5.104).

It is true that in this example we did not have to look very far ahead to discover the additional assumptions that proved useful, but one cannot always expect to be this lucky. The procedure we adopt in cases such as this is to make any assumption we need at the step where we see the need, and to get rid of it when it is no longer needed, and certainly before the last step of the demonstration. Thus, to demonstrate (5.103):

1. $(\exists x)[F(x) \vee G(x)]$	1. Assumption
2. $F(z) \vee G(z)$	2. IE, Step 1
3. $F(z)$	3. Assumption
4. $(\exists x)F(x)$	4. PGE, Step 3
5. $(\exists x)F(x) \to (\exists x)F(x) \vee (\exists y)G(y)$	5. Valid statement formula, $[P \to P \vee Q]$
6. $(\exists x)F(x) \vee (\exists y)G(y)$	6. Mod pon [4,5]
7. $F(z) \to (\exists x)F(x) \vee (\exists y)G(y)$	7. Ded prin [3,6]
8. $G(z)$	8. Assumption
9. $(\exists y)G(y)$	9. PGE, Step 8
10. $(\exists y)G(y) \to (\exists x)F(x) \vee (\exists y)G(y)$	10. Valid statement formula, $[P \to Q \vee P]$
11. $(\exists x)F(x) \vee (\exists y)G(y)$	11. Mod pon [9,10]
12. $G(z) \to (\exists x)F(x) \vee (\exists y)G(y)$	12. Ded prin [8,11]
13. $F(z) \vee G(z) \to (\exists x)F(x) \vee (\exists y)G(y)$	13. Inf by cases [7,12]
14. $(\exists x)F(x) \vee (\exists y)G(y)$	14. Mod pon [2,13]
15. $(\exists x)[F(x) \vee G(x)] \to (\exists x)F(x) \vee (\exists y)G(y)$	15. Ded prin [1,14]

Observe that in this demonstration, Steps 6, 11, and 14 are the same formula. If, however, one stops with Step 6, then the six steps are a proof of (5.107) rather than of (5.103) or of (5.104), since Step 6 is in the "scope" of the assumption at Step 3. The "scope" of an assumption in a demonstration is just the sequence of steps starting with the assumption, and ending with the step just before the assumption is "discharged" by the deduction principle or by some other means.

Let us denote by "S_1" the formula of Step 1; by "S_2" the formula of Step 2, and so on. Then the following exhibits just what is proved at each step of the demonstration:

$S_1 \vdash S_1$
$S_1 \vdash S_2$
$S_1, S_3 \vdash S_3$
$S_1, S_3 \vdash S_4$
$S_1, S_3 \vdash S_5$
$S_1, S_3 \vdash S_6$ [Establishes (5.107)]
$S_1 \vdash S_7$ [Ded prin used here discharges assumption "S_3" from left of $\vdash$.]
$S_1, S_8 \vdash S_8$
$S_1, S_8 \vdash S_9$
$S_1, S_8 \vdash S_{10}$
$S_1, S_8 \vdash S_{11}$ [Establishes (5.108)]
$S_1 \vdash S_{12}$ [Ded prin used here discharges assumption "S_8" from left of $\vdash$.]
$S_1 \vdash S_{13}$
$S_1 \vdash S_{14}$ [Establishes (5.104)]
$\vdash S_{15}$ [Establishes (5.103); ded prin used here discharges assumption "S_1" from left of $\vdash$.]

From this display the scope of the three assumptions is evident. The scope of "S_1" extends from "S_1" to "S_{14}", and thus contains the scopes of "S_3" and "S_8". The scopes of different assumptions may overlap in various ways. One must exercise care in long demonstrations not to use any formula in the scope of an assumption "A" in an inference at some later step after "A" is supposed to have been discharged. To do this is to reassume "A" tacitly at this later step. The only exception is in a case such as "S_5" or "S_{10}". "S_5" is technically in the scope of "S_3", but "S_5" is a valid statement formula, and thus independent of the assumption "S_3". Occasionally one will try to save time by using a formula in the scope of an assumption where the formula is independent of the assumption. To do so is technically a violation of the rules of demonstration, but is an acceptable abbreviation if used with care.

In the course of a demonstration, it is sometimes convenient to establish a formula by introducing its negative as an assumption and showing that this leads to a contradiction. The deduction of a contradiction will be considered sufficient to discharge the assumption and prove the formula. The following demonstration will illustrate the form to be used.

(5.109) $\sim(\exists x)\sim F(x) \rightarrow (x)F(x)$†

→1. $\sim(\exists x)\sim F(x)$	1. Assumption
We establish $F(y)$ indirectly by assuming:	
→2. $\sim F(y)$	2. Assumption
3. $(\exists x)\sim F(x)$	3. PGE, Step 2
4. $(\exists x)\sim F(x) \rightarrow [\sim(\exists x)\sim F(x) \rightarrow F(y)]$	4. Valid statement formula, $[P \rightarrow (\sim P \rightarrow R)]$
5. $\sim(\exists x)\sim F(x) \rightarrow F(y)$	5. Mod pon [3,4]
6. $F(y)$	6. Mod pon [1,5]
7. $F(y)\sim F(y)$	7. Conj inf [6,2]
8. $F(y)$	8. Indirect proof, Steps 2 to 7, [the contradiction discharges the assumption in Step 2]
9. $(x)F(x)$	9. PGU, Step 8
10. $\sim(\exists x)\sim F(x) \rightarrow (x)F(x)$	10. Ded prin [1,9]

Some authors use the device of arrows pointing at the assumption, with the tail passing under the last statement in the scope of the assumption, as indicated in the foregoing demonstration. Sometimes the arrowheads and horizontal tails are omitted and just vertical lines of the proper length are used to indicate the scope of an assumption. Among other devices used might be mentioned the system of successive indentation to indicate scope. For example, the demonstration of (5.103) might be written:

1. $(\exists x)[F(x)\vee G(x)]$
2. $F(z)\vee G(z)$
    3. $F(z)$
    4. $(\exists x)F(x)$
  5. $(\exists x)F(x) \rightarrow (\exists x)F(x)\vee(\exists y)G(y)$
    6. $(\exists x)F(x)\vee(\exists y)G(y)$

† Statement (5.109), and other statements like it, were assumed in Sec. 5.4. Here we prove (5.109) without appealing to any of the assumptions about quantifier conversion.

7. $F(z) \rightarrow (\exists x)F(x) \vee (\exists y)G(y)$
8. $G(z)$
9. $(\exists y)G(y)$
10. $(\exists y)G(y) \rightarrow (\exists x)F(x) \vee (\exists y)G(y)$
11. $(\exists x)F(x) \vee (\exists y)G(y)$
12. $G(z) \rightarrow (\exists x)F(x) \vee (\exists y)G(y)$
13. $F(z) \vee G(z) \rightarrow (\exists x)F(x) \vee (\exists y)G(y)$
14. $(\exists x)F(x) \vee (\exists y)G(y)$
15. $(\exists x)[F(x) \vee G(x)] \rightarrow (\exists x)F(x) \vee (\exists y)G(y)$

We shall not employ any of these devices since in no case do we give a demonstration without an accompanying analysis column. It is easy enough to check the analysis to verify that every assumption is discharged properly by the deduction principle, or indirect proof. Before using some earlier step in a later inference, one should check to see that the earlier step is not in the scope of an assumption that is discharged prior to the inference (the exception to this rule has already been noted).

VI: APPLICATIONS OF LOGIC IN MATHEMATICS

6.1 INTRODUCTION

Associated with the development of each new piece of formal machinery in the preceding chapters were applications in elementary mathematics. In most cases, the available formalism was not entirely adequate, and it was necessary for the success of the illustration to rely on a reader's good sense and experience with mathematics. The reader, having come with us this far, can easily detect the many formal gaps in earlier chapters, where it was necessary to rely on meaning and interpretation. But the formalism now available provides a precise, flexible language for discourse about elementary mathematics. Quite rigorous treatment of an axiomatic system is now possible, enabling us to make a sharp distinction between an abstract theory on the one hand and interpretations of it on the other hand.

The present chapter's concern is mainly with elementary algebra, partly because geometry has already received quite a bit of attention and partly because a rigorous development of geometry is formidable. A rigorous treatment of the miniature geometry, first presented informally in Chap. IV, is given in an appendix. The treatment presented there requires some additional formalism and is complicated.

In Sec. 6.2 we present an axiomatic development of the notion of a *group*. This abstract system deals with the behavior of a binary operation defined on a set of objects; it has interpretations in the addition of real numbers, or in the multiplication of real numbers, but not both at the same time. Later sections present an axiomatic development of the notion of a *field*, which deals with the behavior of two binary operations defined on a single set of objects. The resulting theory has an interpretation in the ordinary arithmetic of real numbers, and thus applies to a large part of elementary algebra.

It is well to iterate the reasons for formal developments such as those to follow. Most of the theorems proved are so familiar that they may seem trivial, particularly if the real-number interpretation is kept in mind. Long symbolic proofs are not presented to convince anyone of the truth of the statements proved. They are presented to show that the theorems are logical consequences of a relatively small set of axioms and that algebra has the same sort of axiomatic structure as has geometry. We suggest that the reader review the last three paragraphs of Sec. 1.1 before embarking on the remainder of this chapter.

Now that formal proofs are understood, we shall feel free to abbreviate formal procedures (with due warning). Informality that consists of conscious, knowledgeable abbreviation of formal steps is quite different from informality that consists of an appeal to meaning, intuition, or an interpretation. In what follows, we shall try to make it clear which kind of informality is being used.

There may be some tendency to feel that abbreviation represents a step backward. We remember a freshman in Fine Arts, who happily followed our formal development of algebra, but grew noticeably unhappy when abbreviations became a practical necessity. On being questioned, she was at first reluctant to criticize, but with some encouragement burst out, "Well! Mathematics is so beautiful that I hate to see it get sloppy like this. I like to see all the steps". We have always been grateful to her for this remark, and regard it as reinforcing our belief that formal treatment of mathematics, properly blended with heuristic treatment, contributes significantly to appreciation of the subject.

6.2 ABSTRACT GROUPS

The first step in building a theory of abstract groups is to add two undefined, nonlogical symbols to the logic. The first is "e", which is the symbol for an individual constant. It is not possible to quantify, or to generalize on e. Such expressions as "$(e)F(e)$" or "$(\exists e)G(e)$" will not appear. It will not be possible to get from "$(\exists x)F(x)$" to "$F(e)$" by IE alone. It will always be possible to get from "$(x)F(x)$" to "$F(e)$" by IU.

The second nonlogical symbol is "$\circ$", which is the symbol for a binary operation. The symbol "$\circ$" is used to produce *terms* such as

$$x \circ y,\ x \circ x,\ (x \circ y) \circ z,\ ((x \circ y) \circ x) \circ z,\ e \circ e,\ e \circ x, \text{ etc.}$$

A term can be recognized as such by a finite number of applications of the following criteria:

a. An individual variable (or constant) is a term.
b. If t_1 and t_2 are terms, then $t_1 \circ t_2$ is a term.

For any interpretation of the abstract system we are about to build, there must be some set of objects G to act as values for variables, and some operator $*$ defined on G to correspond to $\circ$. Further, if under the interpretation, "x" corresponds to an object a and "y" corresponds to an object b, then "$x \circ y$" must correspond to $a*b$, and $a*b$ must be an object in G. This suggests that terms in the logic act like individual variables in some sense. In order to handle the projected theory, the inference rules IU (Sec. 5.16), IS and ISC (Sec. 5.18) must be modified to accommodate terms:

(6.1) IU modified:

$$\frac{(x)F(x)}{F(t)},$$

where t is a term.

(6.2) IS modified:

$$\frac{t_1 = t_2,\ F(t_1)}{F^*(t_2)},$$

where t_1 and t_2 are terms.

ISC is modified similarly, but no other inference rules will be so modified. The extension of inference rules to accommodate terms is necessary for a systematic treatment of algebra, as was foreshadowed in Sec. 5.17. We shall refer to (6.1) and (6.2) by the old notation "IU" and "IS", respectively.

The second step in building the abstract theory is to add five nonlogical axioms to the logic:

Group Axioms:

G_1: $(x,y)(\exists z)[x \circ y = z]$
G_2: $(x,y,z)[(x \circ y) \circ z = x \circ (y \circ z)]$
G_3: $(x,y)[x \circ y = y \circ x]$
G_4: $(x)[x \circ e = x]$
G_5: $(x)(\exists y)[x \circ y = e]$

Since there is only one operational symbol involved in the system, it will be convenient to indicate the operation "$x \circ y$" by simply writing "xy". This is quite similar to the abbreviation of multiplication used in algebra. With this device the axioms are written

$$
\begin{aligned}
&G_1\colon (x,y)(\exists z)[xy = z] \\
&G_2\colon (x,y,z)[(xy)z = x(yz)] \\
(6.3) \qquad &G_3\colon (x,y)[xy = yx] \\
&G_4\colon (x)[xe = x] \\
&G_5\colon (x)(\exists y)[xy = e]
\end{aligned}
$$

Two familiar interpretations of the abstract system are given by

(6.4) Let G be the set of all integers. Let ordinary addition be the binary operation in G. Interpret "e" as another name for zero.

(6.5) Let G be the set of all positive real numbers. Let ordinary multiplication be the binary operation in G. Interpret "e" as another name for 1.

In either of these interpretations, the axioms become true statements in the interpretation. For (6.4), the axioms become the following five statements, which are easily recognized as true statements about the integers:

1. $(x,y)(\exists z)[x + y = z]$ (Closure law)
2. $(x,y,z)[(x + y) + z = x + (y + z)]$ (Associative law)
3. $(x,y)[x + y = y + z]$ (Commutative law)
4. $(x)[x + 0 = x]$
5. $(x)(\exists y)[x + y = 0]$

In the case of (6.5), the axioms become the statements:

1. $(x,y)(\exists z)[xy = z]$ (Closure law)
2. $(x,y,z)[(xy)z = x(yz)]$ (Associative law)
3. $(x,y)[xy = yx]$ (Commutative law)
4. $(x)[x \cdot 1 = x]$
5. $(x)(\exists y)[xy = 1]$

Observe that in the interpretation (6.5), the set G cannot contain the real number 0, since 0 has no reciprocal. Indeed, there is no real number x such that $0 \cdot x = 1$.

Any interpretation of the abstract system is called an *Abelian group*. If the commutativity axiom G_3 is omitted from the system, then any interpretation of the resulting system is called a *Group*.

EXERCISES

Examine the following sets of objects informally to see (*a*) which ones form groups and (*b*) which ones are Abelian groups:

1. A set consisting of just one element.

2. The set of real numbers with ordinary multiplication as the binary operation.

3. Negative real numbers with ordinary multiplication as the binary operation.

4. The integers with ordinary addition as the binary operation.

5. Integers except zero with ordinary multiplication as the binary operation.

6. The real numbers except zero with ordinary multiplication as the binary operation.

7. The set of translations in a plane with one translation following another as the binary operation.

8. The set of rigid motions (translations and rotations) in a plane with one rigid motion following another as the binary operation.

9. The set of numbers 1, 2, 3 with the binary operation defined by the table

∘	1	2	3
1	2	3	1
2	3	1	2
3	1	2	3

10. The set of numbers 1, 2, 3 with the binary operation defined by the table

∘	1	2	3
1	1	2	3
2	2	1	2
3	3	2	1

11. The set of rotations in a plane of an equilateral triangle ABC into itself, with one rotation followed by another as the binary operation.

12. The set of all powers of 2 with the binary operation ordinary multiplication.

The third step in building the abstract group theory is to discover and demonstrate the lemmas and theorems that are logical consequences of steps one and two. We shall adopt an abbreviation in the application of IU and IE to axioms and previously proved theorems. For example, in deriving "$ye = y$" by IU from G_4, we shall omit G_4 as a step and merely write the result "$ye = y$". Then in the analysis column will appear: "IU, G_4".

L_1: $(x)[ex = x]$

1. $xe = x$	1. IU, G_4
2. $xe = ex$	2. IU, G_3
3. $ex = x$	3. IS [2,1]
4. $(x)[ex = x]$	4. PGU

L_1 can be regarded as a commutative form of G_4.† It is a great convenience to have L_1, since in all subsequent proofs, it will now be possible to use it to save three steps. For most of the theorems that follow, we shall follow the proof with an unproved commutative form of the theorem as a corollary.

T_1: $(x,y,z)[x = y \to xz = yz]$

1. $x = y$	1. Assumption
2. $xz = xz$	2. IU (as modified) from R
3. $xz = yz$	3. IS [1,2]
4. $x = y \to xz = yz$	4. Ded prin
5. $(x,y,z)[x = y \to xz = yz]$	5. PGU

Cor: $(x,y,z)[x = y \to zx = zy]$

T_2: $(x,y,z)[xz = yz \to x = y]$ (cancellation)

1. $xz = yz$	1. Assumption
2. $(\exists y)[zy = e]$	2. IU, G_5
3. $zw = e$	3. IE
4. $xz = yz \to (xz)w = (yz)w$	4. IU, T_1
5. $(xz)w = (yz)w$	5. Mod pon [1,4]
6. $(xz)w = x(zw)$	6. IU, G_2
7. $(yz)w = y(zw)$	7. IU, G_2
8. $x(zw) = (yz)w$	8. IS [6,5]
9. $x(zw) = y(zw)$	9. IS [8,7]
10. $xe = ye$	10. IS [3,9]
11. $xe = x$	11. IU, G_4
12. $x = ye$	12. IS [11,10]
13. $ye = y$	13. IU, G_4
14. $x = y$	14. IS [13,12]
15. $xz = yz \to x = y$	15. Ded prin [1,14]
16. $(x,y,z)[xz = yz \to x = y]$	16. PGU

Cor: $(x,y,z)[zx = zy \to x = y]$

† The reader will notice that we have abandoned the agreement made in Sec. 1.4 never to use a letter as the name of a statement. From now on, we shall use the symbol "G_4", for example, both as a symbolic translation of the axiom, and as a name for the axiom, and will no longer make a distinction by means of quotation marks.

With the help of the cancellation theorem, G_5 can be strengthened by showing that for each "x" there is not more than one "y" such that "$xy = e$".

T_3: $(x,y,z)[(xy = e)(xz = e) \rightarrow y = z]$

1. $(xy = e)(xz = e)$	1. Assumption
2. $xy = e$	2. Simp, Step 1
3. $xz = e$	3. Simp, Step 1
4. $xy = xz$	4. IS [2,3]
5. $xy = xz \rightarrow y = z$	5. IU, T_2 cor.
6. $y = z$	6. Mod pon [4,5]
7. $(xy = e)(xz = e) \rightarrow y = z$	7. Ded prin
8. T_3	8. PGU

Cor: $(x,y,z)[(yx = e)(zx = e) \rightarrow y = z]$

T_2 and T_3 together show that to each "x" there exists a unique "y" such that "$xy = e$". Thus it makes sense to give a name to this uniquely determined object by the following definition:

(6.6) $$(x,y)[y = x^{-1} \leftrightarrow xy = e].$$

One may regard "${}^{-1}$" as a singulary operation symbol that produces the term "x^{-1}" when applied to "x". The term "x^{-1}" is called the *inverse* of "x". It follows easily that

L_2: $(x)[xx^{-1} = e]$

1. $(x,y)[y = x^{-1} \leftrightarrow xy = e]$	1. Definition (6.6)
2. $x^{-1} = x^{-1} \leftrightarrow xx^{-1} = e$	2. IU, Step 1
3. $x^{-1} = x^{-1}$	3. IU, R
4. $xx^{-1} = e$	4. Mod pon [3,2]
5. L_2	5. PGU

Cor: $(x)[x^{-1}x = e]$

The next theorem shows that the individual constant e is unique in the sense that no other term has the property expressed in G_4.

T_4: $(x,y)[xy = x \rightarrow y = e]$

1. $xy = x$	1. Assumption
2. $xe = x$	2. IU, G_4
3. $xy = xe$	3. IS [2,1]
4. $xy = xe \rightarrow y = e$	4. IU, T_2 cor.
5. $y = e$	5. Mod pon [3,4]
6. $xy = x \rightarrow y = e$	6. Ded prin
7. T_4	7. PGU

The next theorem shows that it is always possible to find a solution for an equation "$ax = b$" in an Abelian group.

T_5: $(x,z)(\exists y)[xy = z]$

1. $ez = z$	1. IU, L_1
2. $xx^{-1} = e$	2. IU, L_2
3. $(xx^{-1})z = z$	3. IS [1,2]
4. $(xx^{-1})z = x(x^{-1}z)$	4. IU, G_2
5. $x(x^{-1}z) = z$	5. IS [4,3]
6. $(x,y)(\exists w)[xy = w]$	6. G_1
7. $(\exists w)[x^{-1}z = w]$	7. IU, Step 6
8. $x^{-1}z = y$	8. IE, Step 7
9. $xy = z$	9. IS [8,5]
10. $(\exists y)[xy = z]$	10. PGE
11. T_5	11. PGU.

EXERCISES

Write out a proof for each of the following statements:

1. $(y,b)(\exists a)[ay = b]$
2. Without use of G_3: $(x)[xe = ex]$
3. $(x)[(x^{-1})^{-1} = x]$
4. $(x,y)[x^{-1}y^{-1} = (yx)^{-1}]$
5. $(x)[x \neq e \rightarrow x^{-1} \neq e]$
6. $(x,y)[xy^{-1} = (yx^{-1})^{-1}]$
7. $(x,y)[x(y^{-1})^{-1} = xy]$

6.3 ISOMORPHIC INTERPRETATIONS

The concept of an Abelian group[1] is important to many branches of mathematics—algebra and topology, to name two. The Abelian groups that occur in the various branches of mathematics are interpretations of the foregoing abstract system. The development of the abstract system can be thought of as an investigation into the behavior of an arbitrary set of elements on which is defined a binary operation having certain natural properties. It is typical of modern mathematics to study such systems in the abstract, rather than via interpretations having additional properties, but properties that are irrelevant. In other words, if it is the behavior of binary operations that is of interest, then instead of investi-

[1] For a good and accessible introduction to groups, see *The Twenty-third Yearbook of the National Council of Teachers of Mathematics*, "Insights Into Modern Mathematics," chap. V, Washington, D.C.: National Council of Teachers of Mathematics, 1957.

gating real numbers under addition, or rational numbers under multiplication, polynomials under addition or other existing structures, one strips away all irrelevant properties and attempts to see what happens in a system having just the properties of interest.

The investigation may be broadened by dropping some of the axioms, or adding some, or both. If axiom G_3 is dropped, the resulting structure is that of a *group*. If G_5 is dropped, the resulting structure is that of a *semigroup*. Group theory is often divided into the theory of *finite* and *infinite* groups. If a notion of limits is added to the group structure, the result is a theory of *continuous groups*. Abstract development of these notions is valuable because in a single investigation theorems are discovered and proved that have valid interpretations in many different branches of mathematics.

We have given two important examples of infinite Abelian groups in (6.4) and (6.5). We next give two examples of finite Abelian groups, and then take up the question: When are two interpretations essentially different?

(6.7) Let G be the set of elements $\{1,2,3,4,5,6\}$ with the binary operation taken to be *multiplication modulo* 7.

To multiply x and y modulo 7, form the product xy under ordinary multiplication. Then divide xy by 7, and take the remainder to be the product modulo 7. That is,

$$3 \cdot 4 = 5 \pmod 7$$

because $3 \cdot 4 = 12$ by ordinary multiplication, and on division of 12 by 7 the remainder is 5.

To verify that G is an Abelian group, it is necessary to check that the group axioms, considered as statements about multiplication mod 7 of the elements of G, are true. That multiplication mod 7 is defined for every pair of elements of G, and is associative and commutative follows easily from the corresponding properties of ordinary multiplication of positive integers. It is easy to see that 1 is the unit element (corresponding to e) of the group. To check for inverses, one observes, for example,

$$2 \cdot 4 = 1.$$

Hence, $2^{-1} = 4$, and also, $4^{-1} = 2$. The complete table of inverses follows, but it should be checked:

$$\begin{array}{ll} 1^{-1} = 1 & \quad 4^{-1} = 2 \\ 2^{-1} = 4 & \quad 5^{-1} = 3 \\ 3^{-1} = 5 & \quad 6^{-1} = 6 \end{array}$$

(6.8) Let H be the set of elements $\{0,1,2,3,4,5\}$ with the binary operation *addition modulo* 6.

To add mod 6, add in the usual way, then reduce the answer mod 6. It will probably be convenient here to symbolize the group operation by "$+$", and to symbolize the inverse of "x" by "$(-x)$". It is easy to see that "$(x)[x + 0 = x]$" and hence that the unit for this group is 0. The rest of the structure should be checked as in the previous example.

EXERCISES

1. A set of three objects can be ordered in six different ways. Any particular ordering is called an *arrangement.* The change from one arrangement to another is called a *permutation.* A permutation of three objects can be represented by

$$\begin{pmatrix} 1 & 2 & 3 \\ 3 & 2 & 1 \end{pmatrix}.$$

The numerals in this symbol refer to positions. The top row refers to the positions of the three objects before permutation; the bottom row refers to the positions the objects end up in after permutation. Thus, this permutation carries the object in the first position to the third position, leaves the object in second position fixed, and carries the object in third position to the first position. This permutation could also be written

$$\begin{pmatrix} 2 & 3 & 1 \\ 2 & 1 & 3 \end{pmatrix} \quad \text{or} \quad \begin{pmatrix} 3 & 1 & 2 \\ 1 & 3 & 2 \end{pmatrix},$$

and in various other ways.

Let G be the set of all permutations on three objects. If one permutation is followed by another, the resulting effect can always be produced by a single permutation which is defined as the *product* of the two permutations.

a. Write out the six permutations of G.
b. Work out a scheme for finding the product of any two permutations, and check it by application to specific arrangements.
c. Which element plays the role of e?
d. Write out the inverse of each element.
e. Make up a multiplication table for the group. (Observe that the group is not Abelian.)
f. Find as many subgroups of G as you can. (A subgroup of G is a subset of elements H forming a group under the operation of G.)
g. Investigate the group of permutations on four objects.

2. Let G be the set {A,B,C}. The operation is defined by the following table:

$\circ$	A	B	C
A	A	B	C
B	B	C	A
C	C	A	B

a. Which member plays the role of e?
b. Write the inverse of each element.

3. Let G be the set $\{1,-1,i,-i\}$, where $i^2 = -1$. The operation is ordinary multiplication.
a. Which element plays the role of e?
b. Form a multiplication table for the group.
c. Write the inverse of each element.

We now consider the question: Are the Abelian groups G (6.7) and H (6.8) essentially different?

Any two interpretations of the same abstract system are "the same" in the sense that they both have all the properties expressed in the axioms of the abstract system. Every interpretation, however, has properties in addition to those of the defining system; hence, two interpretations can differ, and indeed, must differ in some sense in order to be distinguishable. We shall say that two interpretations are *isomorphic* if there exists a *one-to-one correspondence* between their respective elements that *preserves the structures* of the interpretations. We shall take up the meaning of the italicized phrases in this informal statement for the case of two Abelian groups, G and G^*.

To say that there exists a one-to-one correspondence between G and G^* means simply that to every $x \varepsilon G$ there exists a unique corresponding element $\phi(x) \varepsilon G^*$, and if $x \varepsilon G$ and $y \varepsilon G$ correspond to the same element of G^* [that is, if $\phi(x) = \phi(y)$], then $x = y$.† Finally, for every $x^* \varepsilon G^*$ there is an $x \varepsilon G$ such that $\phi(x) = x^*$. Hence, under the correspondence ϕ, each element of G corresponds to a unique element of G^*, and each element of G^* corresponds to a unique element of G. (Finite groups having the same number of elements can always be paired in a 1-1 correspondence. For infinite groups, the matter is not so obvious.)

The structure of an Abelian group is determined by the multiplication table for its binary operation, the relation of each element to its inverse, and the properties of its unit element. To say that a correspondence preserves the binary operations is to say that if x (an element of G) cor-

† Recall that "$x \varepsilon G$" means: "x is an element of G".

responds to x^* (an element of G^*), and y corresponds to y^*, then the product xy corresponds to the product x^*y^*, and this must hold for every possible choice for the elements x and y. Using the function symbol "ϕ", we can express the property of preserving the binary operations neatly by

$$(x,y)[\phi(xy) = \phi(x)\phi(y)],$$

where it is understood that "x" and "y" name elements of G, and "$\phi(x)$" and "$\phi(y)$" name corresponding elements of G^*. For the groups to be isomorphic, it must also be the case that

(6.9)
a. $\phi(e) = e^*$, where e and e^* are unit elements of G and G^*, respectively.
b. $(x)\{\phi(x^{-1}) = [\phi(x)]^{-1}\}$, i.e., inverses correspond.

It can be proved that if the binary operations are preserved, then (6.9*a*) and (6.9*b*) also hold. Hence, isomorphism of groups is usually defined as follows:

Definition of Group Isomorphism (holds for groups as well as Abelian groups):

(6.10) There is a correspondence ϕ between the elements of G and G^* such that
a. ϕ is 1-1.
b. $(x,y)[\phi(xy) = \phi(x)\phi(y)]$.

If G and G^* are isomorphic, we write $G \cong G^*$.

As an example of isomorphic groups, we show that the group

$$G = \{1,2,3, \ldots ,6\}$$

with multiplication modulo 7, and the group $G^* = \{\mathbf{0}, \mathbf{1}, \ldots, \mathbf{5}\}$ with addition modulo 6, are isomorphic. We show the correspondence schematically:

Diagram of Correspondence

G	G^*
1	**0**
2	**1**
3	**2**
4	**3**
5	**4**
6	**5**

and in terms of the functional ϕ, the elements $\phi(x)$ of G^* are:

Table for ϕ

$\phi(1) = \mathbf{0}$
$\phi(2) = \mathbf{2}$
$\phi(3) = \mathbf{1}$
$\phi(4) = \mathbf{4}$
$\phi(5) = \mathbf{5}$
$\phi(6) = \mathbf{3}$

To help distinguish the groups, we symbolize the group operation in G by "$\cdot$" and the operation in G^* by "$+$". Then, $\phi(1) = \mathbf{0}$ shows that the unit elements correspond. We test preservation of inverses below (we indicate the inverse of x^* in G^* by $-x^*$):

Table of inverses, G	*Table of inverses, G**
$1^{-1} = 1$	$-\mathbf{0} = \mathbf{0}$
$2^{-1} = 4$	$-\mathbf{1} = \mathbf{5}$
$3^{-1} = 5$	$-\mathbf{2} = \mathbf{4}$
$4^{-1} = 2$	$-\mathbf{3} = \mathbf{3}$
$5^{-1} = 3$	$-\mathbf{4} = \mathbf{2}$
$6^{-1} = 6$	$-\mathbf{5} = \mathbf{1}$

Then

$$\begin{array}{llll}
\phi(1^{-1}) = \phi(1) = \mathbf{0} = -\mathbf{0} = -\phi(1) & \text{or} & \phi(1^{-1}) = -\phi(1) \\
\phi(2^{-1}) = \phi(4) = \mathbf{4} = -\mathbf{2} = -\phi(2) & \text{or} & \phi(2^{-1}) = -\phi(2) \\
\phi(3^{-1}) = \phi(5) = \mathbf{5} = -\mathbf{1} = -\phi(3) & \text{or} & \phi(3^{-1}) = -\phi(3) \\
\phi(4^{-1}) = \phi(2) = \mathbf{2} = -\mathbf{4} = -\phi(4) & \text{or} & \phi(4^{-1}) = -\phi(4) \\
\phi(5^{-1}) = \phi(3) = \mathbf{1} = -\mathbf{5} = -\phi(5) & \text{or} & \phi(5^{-1}) = -\phi(5) \\
\phi(6^{-1}) = \phi(6) = \mathbf{3} = -\mathbf{3} = -\phi(6) & \text{or} & \phi(6^{-1}) = -\phi(6)
\end{array}$$

So we have

$$(x)[\phi(x^{-1}) = -\phi(x)],$$

and inverses correspond.

To check finally that

$$(x,y)[\phi(xy) = \phi(x) + \phi(y)]$$

we proceed to check, for example, products formed with 2 (of G) as one factor:

$$\begin{array}{l}
\phi(2 \cdot 1) = \phi(2) = \mathbf{2} = \mathbf{2} + \mathbf{0} = \phi(2) + \phi(1) \\
\phi(2 \cdot 2) = \phi(4) = \mathbf{4} = \mathbf{2} + \mathbf{2} = \phi(2) + \phi(2) \\
\phi(2 \cdot 3) = \phi(6) = \mathbf{3} = \mathbf{2} + \mathbf{1} = \phi(2) + \phi(3) \\
\phi(2 \cdot 4) = \phi(1) = \mathbf{0} = \mathbf{2} + \mathbf{4} = \phi(2) + \phi(4) \\
\phi(2 \cdot 5) = \phi(3) = \mathbf{1} = \mathbf{2} + \mathbf{5} = \phi(2) + \phi(5) \\
\phi(2 \cdot 6) = \phi(5) = \mathbf{5} = \mathbf{2} + \mathbf{3} = \phi(2) + \phi(6)
\end{array}$$

The remaining products will be found to check similarly.

There are thirty-six different ways of setting up a one-one correspondence between the elements of G and G^*. Of these 36 correspondences, or functions, some have the properties (6.10*b*) and some do not. It turns out that any function that does not make the unit elements correspond does not have the property (6.10*b*). Also, a function that makes 2 in G correspond to **1** in G^* will not have the property (6.10*b*). However, the function ψ defined by the following table does have the required property:

$$\psi(1) = \mathbf{0} \quad \psi(2) = \mathbf{4} \quad \psi(3) = \mathbf{5} \quad \psi(4) = \mathbf{2} \quad \psi(5) = \mathbf{1} \quad \psi(6) = \mathbf{3}$$

Although it turns out that the functions ϕ and ψ are the only ones having properties (6.10) for these groups, remember that the existence of one such function is enough to ensure isomorphism.

Now we prove that if ϕ has property (6.10*b*), all the structure is preserved.

T_6: If ϕ is a correspondence between the elements of a group G and the elements of a group G^*, such that

$$(x,y)[\phi(xy) = \phi(x)\phi(y)],$$

then

$$a.\ \phi(e) = e^*$$
$$b.\ (x)\{\phi(x^{-1}) = [\phi(x)]^{-1}\}.$$

Proof of (*a*):

1. $(xy)[\phi(xy) = \phi(x)\phi(y)]$	1. Assumption
2. $\phi(a) = \phi(a)$	2. IU, R
3. $ae = a$	3. IU, G_4
4. $\phi(a) = \phi(ae)$	4. IS [3,2]
5. $\phi(ae) = \phi(a)\phi(e)$	5. IU, Step 1
6. $\phi(a)\phi(e) = \phi(a)$	6. IS [5,4]
7. $\phi(a)\phi(e) = \phi(a) \rightarrow \phi(e) = e^*$	7. IU, T_4 (substitution instance: $\phi(a)$ for x; $\phi(e)$ for y)
8. $\phi(e) = e^*$	8. Mod pon [6,7]
9. $(x,y)[\phi(xy) = \phi(x)\phi(y)] \rightarrow \phi(e) = e^*$	9. Ded prin

Proof of (*b*):

1. $(xy)[\phi(xy) = \phi(x)\phi(y)]$	1. Assumption
2. $\phi(aa^{-1}) = \phi(aa^{-1})$	2. IU, R
3. $aa^{-1} = e$	3. IU, L_2
4. $\phi(e) = \phi(aa^{-1})$	4. IS [3,2]
5. $\phi(aa^{-1}) = \phi(a)\phi(a^{-1})$	5. IU, Step 1
6. $(x,y)[\phi(xy) = \phi(x)\phi(y)] \rightarrow \phi(e) = e^*$	6. T_6 (*a*)
7. $\phi(e) = e^*$	7. Mod pon [1,6]
8. $e^* = \phi(aa^{-1})$	8. IS [7,4]
9. $\phi(a)\phi(a^{-1}) = e^*$	9. IS [8,5]
10. $\phi(a)\phi(a^{-1}) = e^* \rightarrow \phi(a^{-1}) = [\phi(a)]^{-1}$	10. IU, Definition (6.6)
11. $\phi(a^{-1}) = [\phi(a)]^{-1}$	11. Mod pon [9,10]
12. $(x)\{\phi(x^{-1}) = [\phi(x)]^{-1}\}$	12. PGU
13. $(x,y)[\phi(xy) = \phi(x)\phi(y)] \rightarrow (x)\{\phi(x^{-1}) = [\phi(x)]^{-1}\}$	13. Ded prin

Two examples of isomorphic groups follow. Let

G be the set of integers with ordinary addition;
G^* be the set of even integers with ordinary addition.

Then the 1-1 correspondence ϕ defined by the statement

$$(x)[x \varepsilon G \rightarrow \phi(x) = 2x]$$

shows that $G \cong G^*$. For

$$\phi(x + y) = 2(x + y) = 2x + 2y = \phi(x) + \phi(y).$$

The fact that $G \cong G^*$ means that as far as ordinary addition is concerned, the integers and the even integers exhibit the same behavior. Let

G be the set of positive real numbers with ordinary multiplication;
G^* be the set of real numbers with ordinary addition.

Then the existence of the 1-1 correspondence ϕ defined below demonstrates $G \cong G^*$

$$(x)[\phi(x) = \log x].$$

For

$$\phi(xy) = \log (xy) = \log x + \log y = \phi(x) + \phi(y).$$

EXERCISES

Test the systems in each of the following exercises to see if the conditions (6.10a) and (6.10b) for an isomorphism can be satisfied:

1. G: the set of integers with ordinary multiplication
G^*: the set of natural numbers with ordinary addition

2. $G = \{a,b,c\}$ with the binary operation defined by the table:

$\circ$	a	b	c
a	a	b	c
b	b	c	a
c	c	a	b

$G^* = \{1,2,3\}$ with the binary operation defined by the table:

$\otimes$	1	2	3
1	3	1	2
2	1	2	3
3	2	3	1

3. $G = \{2,4,6,8, \ldots\}$ with the binary operation ordinary addition

$G^* = \{1,2,3, \ldots\}$ with the binary operation ordinary addition

4. Is there an isomorphism of the two systems of Exercise 3 if the binary operation in each case is ordinary multiplication?

5. $G = \{1,2,3\}$ with the binary operation defined by the table:

$+$	1	2	3
1	1	2	3
2	2	3	1
3	3	1	2

$G^* = \{a,b,c\}$ with the binary operation defined by the table:

$*$	a	b	c
a	c	b	a
b	b	c	a
c	a	a	c

6. $G = \{a,b,c,d\}$ with the binary operation defined by the table:

$\oplus$	a	b	c	d
a	a	b	c	d
b	b	a	d	c
c	c	d	a	b
d	d	c	b	a

$G^* = \{e,i,a,u\}$ with the binary operation defined by the table:

$\otimes$	e	i	a	u
e	e	i	a	u
i	i	e	u	a
a	a	u	e	i
u	u	a	i	e

6.4 ABSTRACT FIELD SYSTEM

Elementary algebra has a more complex structure than that of the Abelian groups of the foregoing sections, since at least two distinct binary operations are required. Yet the group structure is to be found in elementary algebra, for the real numbers under addition form an Abelian group, and so do the real numbers (except zero) under multiplication. One way to set up an abstract system expressing the laws of algebra is to consider a universe having two distinct binary operations defined on it, with appropriate axioms to produce two Abelian groups. To connect the two groups operationally, it will be necessary to state an axiom concerning combination of the two operations. Interpretations of the resulting system are called *fields*.

The two operations will be symbolized by "+" and "·". The symbols are borrowed from arithmetic and the operations are usually called simply *addition* and *multiplication*, even though they are entirely abstract operations, not to be confused with the operations of the same names in arithmetic and algebra. Two individual constants will be needed, and will be symbolized by "0" and "1". We immediately adopt the abbreviation of writing "xy" for "$x \cdot y$".

Field Axioms:

A_1: $(x,y)(\exists z)[x + y = z]$
A_2: $(x,y,z)[(x + y) + z = x + (y + z)]$
A_3: $(x,y)[x + y = y + x]$
A_4: $(x)[x + 0 = x]$
A_5: $(x)(\exists y)[x + y = 0]$
M_1: $(x,y)(\exists z)[xy = z]$
M_2: $(x,y,z)[(xy)z = x(yz)]$
M_3: $(x,y)[xy = yx]$
M_4: $(x)[x \cdot 1 = x]$
M_5: $(x)\{x \neq 0 \rightarrow (\exists y)[xy = 1]\}$
D_1: $(x,y,z)[x(y + z) = (xy) + (xz)]$
D_2: $1 \neq 0$

The first five axioms are just the axioms for an Abelian group. The next five axioms are also the axioms for an Abelian group, except for the restriction embodied in M_5. Note that in applying IU to any axiom, 1 and 0 are legitimate values for obtaining substitution instances, even in the case of M_5.

It is customary to call axioms A_1 and M_1 *closure axioms;* A_2 and M_2 *associativity axioms;* and A_3 and M_3 *commutativity axioms.* D_1 is called a *distributivity axiom;* it connects the two field operations. We shall make the usual agreements about omission of parentheses in expressions involving both operations (see Sec. 2.9), and shall write "$xy + xz$" for "$(xy) + (xz)$" in what follows.

D_2 is assumed for the purpose of ruling out the trivial interpretation consisting of a single individual. Without D_2, the set consisting of the single element n would be a field if the operations were given by the table:

$$n + n = n$$
$$n \cdot n = n$$
$$1 = n = 0$$

Inverses for the two operations may be defined as in the system for abstract groups. A_5 and M_5 establish existence of additive and multiplicative inverses, and the analogue of T_3, Sec. 6.2, may be proved for each of the field operations to establish uniqueness of inverses. We shall not prove these analogue theorems, but state them as

L_0: *a.* $(x,y,z)[(x + y = 0)(x + z = 0) \rightarrow y = z]$.
b. $(x,y,z)[(xy = 1)(xz = 1) \rightarrow y = z]$.

A_5, M_5, and L_0 justify defining two inverse operators "$-$" and "$^{-1}$" by means of the following equivalences:

Definition of Additive Inverse:

(6.11) $$(x,y)[y = -x \leftrightarrow x + y = 0]$$

Definition of Multiplicative Inverse:[1]

(6.12) $$(x,y)\{x \neq 0 \rightarrow [y = x^{-1} \leftrightarrow xy = 1]\}$$

The next lemma follows immediately. It is an analogue of L_2, Sec. 6.2,

L_1: $(x)[x + (-x) = 0]$

[1] The form of this definition entails some subtle difficulties that we do not pursue here. For a good discussion of the treatment of zero in definitions of division, see Patrick Suppes, *Introduction to Logic,* pp. 163–169. Princeton, N.J.: D. Van Nostrand Company, Inc., 1957.

The proof is left as an exercise. A similar lemma about the multiplicative inverse has a slightly longer proof. We state it without proof:

L_2: $(x)[x \neq 0 \rightarrow xx^{-1} = 1]$

A number of theorems can now be stated without proof since they have already been proved as theorems of the system for abstract groups of the previous section. The proofs for the present system go through in the same way with obvious notational changes.

T_1: $(x,y,z)[x = y \leftrightarrow x + z = y + z]$
Cor: $(x,y,z)[x = y \leftrightarrow z + x = z + y]$

T_2: $(x,y,z)[x = y \rightarrow xz = yz]$
Cor: $(x,y,z)[x = y \rightarrow zx = zy]$

T_3: $(x,y,z)\{z \neq 0 \rightarrow [xz = yz \rightarrow x = y]\}$
Cor: $(x,y,z)\{z \neq 0 \rightarrow [zx = zy \rightarrow x = y]\}$

In fact, the analogue of T_3 does not exist in group theory, but the proof of T_3 is only a slight modification of the proof of T_2, Sec. 6.2, and is left as an exercise. T_3 expresses a cancellation law for multiplication, and T_1 contains a cancellation law for addition.[1]

The next lemmas are useful in dealing with equations of the system.

L_3: $(x,y,z)[x + y = z \rightarrow x = z + (-y)]$

1. $x + y = z$	1. Assumption
2. $x + y = z \rightarrow (x + y) + (-y) = z + (-y)$	2. IU, T_1
3. $(x + y) + (-y) = z + (-y)$	3. Mod pon [1,2]
4. $(x + y) + (-y) = x + (y + (-y))$	4. IU, A_2
5. $x + (y + (-y)) = z + (-y)$	5. IS [4,3]
6. $y + (-y) = 0$	6. IU, L_1
7. $x + 0 = z + (-y)$	7. IS [6,5]
8. $x + 0 = x$	8. IU, A_4
9. $x = z + (-y)$	9. IS [8,7]
10. $x + y = z \rightarrow x = z + (-y)$	10. Ded prin
11. L_3	11. PGU

Cor: $(x,y,z)[y + x = z \rightarrow x = (-y) + z]$

With some slight modification the next lemma is proved similarly:

L_4: $(x,y,z)\{y \neq 0 \rightarrow [xy = z \rightarrow x = zy^{-1}]\}$
Cor: $(x,y,z)\{y \neq 0 \rightarrow [yx = z \rightarrow x = y^{-1}z]\}$

[1] These cancellation laws should not be confused with a student's cancellation law well known to all mathematics teachers which states: "If a symbol appears twice on a page, cancel".

6.5 INTERPRETATIONS OF THE ABSTRACT SYSTEM

An interpretation of the abstract system is a set F of objects having two operations defined on it. The abstract operations correspond with these operations defined on F. There must be two unit elements of F to correspond to the individual constants 1 and 0 of the abstract system. With the correspondences so determined, the axioms can be interpreted as statements about F; if they are true statements about F, F is called a *field*. Two examples follow.

The Field of Real Numbers:

(6.13) Let F be the set of real numbers. Take 0 and 1 as the unit elements, and ordinary addition and multiplication for the binary operations.

All the axioms are recognizable as familiar properties of real numbers. Of course, the real numbers have other properties than those given by the axioms, or deducible from the axioms. For instance, the real numbers are ordered by an order relation ("$<$"), and there are infinitely many real numbers.

Other fields have their own special properties that distinguish them from the field of real numbers. An example is (6.14).

The Field of Integers Modulo 7 (From now on denoted by "F[mod 7]"):

(6.14) Let F be the set of numbers $\{0,1,2,3,4,5,6\}$. Take 0 and 1 as the unit elements. Take for the binary operations addition and multiplication modulo 7.

It is easy to see that field axioms $A_1 \cdots A_5$, $M_1 \cdots M_5$ are satisfied. Clearly D_2 is satisfied; so there remains the check of D_1.

Since there are only a finite number of substitution instances of D_1 in this interpretation, we could check each one. This would entail checking 196 cases. It will be more economical of effort to show that D_1 in the interpretation follows from the distributivity of the real numbers. Since we will need here to distinguish between addition and multiplication of real numbers, and addition and multiplication in our finite field, we distinguish these operations for the sake of this informal proof by using "$\oplus$" and "$\circ$" to denote the operations in the finite field.

Now by definition, $x \circ y$ is the remainder obtained on dividing xy by 7. Hence,

$$xy = 7q + x \circ y$$

for some integer q, and

$$xz = 7r + x \circ z$$

for some integer r. (Observe that both $x \circ y$ and $x \circ z$ are less than 7 by definition of *remainder*.) Then adding,

$$xy + xz = 7(q + r) + x \circ y + x \circ z.$$

But,

$$x \circ y + x \circ z = 7s + x \circ y \oplus x \circ z$$

for some integer s; so

$$xy + xz = 7(q + r + s) + x \circ y \oplus x \circ z.$$

On the other hand,

$$y + z = 7a + y \oplus z$$

for some integer a; and so

$$x(y + z) = 7ax + x(y \oplus z).$$

Now,

$$x(y \oplus z) = 7b + x \circ (y \oplus z).$$

So by substitution

$$x(y + z) = 7(ax + b) + x \circ (y \oplus z).$$

But D_2 holds for ordinary addition and multiplication of real numbers; so

$$x(y + z) = xy + xz.$$

Hence,

$$7(ax + b) + x \circ (y \oplus z) = 7(q + r + s) + x \circ y \oplus x \circ y.$$

Now, both $x \circ (y \oplus z)$ and $x \circ y \oplus x \circ z$ are remainders after division by 7; hence, it follows from the last equation that these remainders are equal. (If two numbers are equal, their remainders on division by 7 are equal.)

This field is finite and hence quite different from the field of real numbers. As will be shown later on, the field of integers modulo 7 cannot be ordered as can the real number field. However, F[mod 7] is a field, and one can expect to solve equations in it (simultaneous systems of equations included) and do many other of the same sorts of things that can be done with the field of real numbers.

6.6 FURTHER DEVELOPMENT OF THE ABSTRACT FIELD SYSTEM

Before proving further theorems of the abstract system, let us reiterate the purpose of these demonstrations. We are trying to show that certain statements in the system are theorems. That is, that the statements are logical consequences of the small set of axioms. As asserted earlier, we are not trying to convince anyone that the statements are true statements about numbers. One needs no convincing that the statements in a real-number interpretation are true. Perhaps conviction is not so certain in the interpretation (6.14), but in any event, we prove the theorems for formal reasons, and conviction or faith is not in question.

T_4: $(x,y,z)[(y + z)x = yx + zx]$

1. $x(y + z) = xy + xz$	1. IU, D_1
2. $x(y + z) = (y + z)x$	2. IU, M_3
3. $(y + z)x = xy + xz$	3. IS [2,1]
4. $xy = yx$	4. IU, M_3
5. $(y + z)x = yx + xz$	5. IS [4,3]
6. $xz = zx$	6. IU, M_3
7. $(y + z)x = yx + zx$	7. IS [6,5]
8. $(x,y,z)[(y + z)x = yx + zx]$	8. PGU

Numerous statements similar to T_4 can be proved as consequences mainly of the associative and commutative axioms, with occasional use of D_1. We give one further example.

T_5: $(x,y,z)[(x + y) + z = (z + x) + y]$

1. $(x + y) + z = x + (y + z)$	1. IU, A_2
2. $y + z = z + y$	2. IU, A_3
3. $(x + y) + z = x + (z + y)$	3. IS [2,1]
4. $(x + z) + y = x + (z + y)$	4. IU, A_2
5. $(x + y) + z = (x + z) + y$	5. IS [4,3]
6. $x + z = z + x$	6. IU, A_3
7. $(x + y) + z = (z + x) + y$	7. IS [6,5]
8. T_5	8. PGU

EXERCISE

1. Show that *subtraction* is not an associative operation
  a. By counter example in the field of real numbers,
  b. By counter example, or other means, for the abstract system.

6.7 TYPES OF INFORMALITY IN PROOFS

The proof of T_5 is somewhat informal, since here, as in the sections on groups, we omit the statements of the axioms as steps in the proof. An unabbreviated demonstration of T_5 would have, preceding our Step 1, a statement of A_2. The degree of informality, or what is the same, the amount of abbreviation, permitted in proofs depends upon the purpose for carrying out the proof. As we pointed out before, a rigorous demonstration has the advantage that it is easy to check for validity since the definition of "demonstration" is relatively simple. A disadvantage of demonstrations is their excessive length. Informal proofs are usually shorter than demonstrations, but accurate description of what is meant by "proof" is much more complicated to state. Furthermore, errors in logic are less likely in demonstrations than in informal proofs. A good compromise is to consider any informal proof to be an abbreviation of a demonstration, where the types of abbreviations to be allowed are specified. The abbreviations may be of the logical forms of proof, as suggested in Sec. 3.10; or abbreviation may consist in omitting mention of certain specified theorems and axioms; or finally, the abbreviation may involve the notation, for example, the agreement to omit parentheses in "$(xy) + (xz)$".

In any rigorous development of algebra from axioms, such as we have begun with the field axioms, we would not go on forever proving theorems as formally as we have. As the theory develops, abbreviations can be introduced systematically. The type of abbreviations and the timing of their introduction are a matter for good pedagogical judgment.

Algebra has a great advantage over geometry as a vehicle for teaching axiomatic structure. For not only is the required logical apparatus simpler, but rigorous demonstrations of the early theorems of algebra are far less lengthy than are demonstrations of the early theorems even of the miniature geometry, as a glance at the Appendix will show. Demonstrations of most theorems of Euclidean geometry are much longer than those of the miniature geometry. As a consequence, abbreviation of proof is much greater in plane geometry, and must come much sooner there than would be necessary in algebra. An important device for shortening proofs in plane geometry is to reason from the figure—while at the same time telling a student that he must not do this. Such devices and other abbreviations of proof are so numerous in plane geometry that it is extremely difficult to give a clear description of what a proof is. If a clear description of *proof* is not available, then no clear description of abbreviation of proofs is possible.

We list typical abbreviations with comments on their use in this chapter.

Logical Abbreviations	*Comment*
1. IU with major premise tacit	1. Full use (Already discussed in this section)
2. Multiple use of IS	2. Generally full use (discussed below)
3. Tacit use of commutativity, i.e., "AB" for "BA"	3. Often used
4. "ABC" for "$(AB)C$" and "$A \vee B \vee C$" for "$(A \vee B) \vee C$"	4. Full use
5. Discharge of an assumption by indirect proof	5. Full use (discussed in Sec. 5.19)

6. Logical substitution from a valid statement formula of equivalence, with the valid statement formula tacit	6. Full use (discussed below)
7. Modus ponens with major premise tacit	7. Rarely used
8. IS with one premise tacit	8. Not used
9. Conjunctive inference tacit	9. Not used (will be used in Appendix)
10. Simplification tacit	10. Not used (will be used in Appendix)

Multiple use of IS can be illustrated by modifying the proof of T_4. Simply omit Steps 3 and 5. Then as the analysis for Step 7, write "IS [2,4,6 in 1]". Omission of any one of the Steps 2, 4, 6 in proving T_4 would be an example of abbreviation of Type 8.

As an illustration of a Type 6 abbreviation consider that at Step 7 of a demonstration we have

7. $(x = y)\vee(x = z)$	7. Assumption

and we want to obtain "$x \neq y \rightarrow x = z$". A full demonstration would call for

7. $(x = y)\vee(x + z)$	7. Assumption
8. $[(x = y)\vee(x = z)] \leftrightarrow [(x \neq y) \rightarrow (x = z)]$	8. Valid statement formula
9. $(x \neq y) \rightarrow (x = z)$	9. Logical substitution (Sec. 2.11) or mod pon

We abbreviate as follows:

7. $(x = y)\vee(x = z)$	7. Assumption
8. $(x \neq y) \rightarrow (x = z)$	8. Substitution, valid statement formula

Field System Abbreviations	*Comment*
1. Omission of parentheses in mixed expressions ("$xy + xz$" for "$(xy) + (xz)$")	1. Full use (already discussed in this section)
2. "$x + y + z$" for "$(x + y) + z$" "xyz" for "$(xy)z$"	2. Used in interpretations, but not in the abstract system (discussed below)
3. Associativity and commutativity tacit	3. Used in interpretations, but not in the abstract system
4. Use of cancellation theorems in form of derived inference rules	4. Not used (discussed below)

The expression "$x + y + z$" is not a term in the system (see Sec. 6.2). By agreement, however, it may be taken as shorthand for "$(x + y) + z$". Extending the agreement, one may write "$x + y + z + w$" for "$[(x + y) + z] + w$". It is then but a short step to the abbreviation of leaving the associative axiom for addition completely tacit. Such abbreviation is quite useful in algebra, and we may use it in interpretations of fields. We shall not use it in the abstract development, however, since to do so would obscure the role of Axiom A_2. Similar remarks are of course possible about multiplication.

As an example of a Type 4 abbreviation, we might write

1. $m \neq 0$	1. Assumption
2. $mx = my$	2. Assumption
3. $x = y$	3. Cancellation

This is an abbreviation for

1. $m \neq 0$	1. Assumption
2. $mx = my$	2. Assumption
3. $m \neq 0 \rightarrow [mx + my \rightarrow x = y]$	3. IU, T_3 Cor.
4. $mx + my \rightarrow x = y$	4. Mod pon [1,3]
5. $x = y$	5. Mod pon [2,4]

In effect, "cancellation" in the abbreviated version indicates use of an inference rule derived from T_3, namely,

$$\frac{m \neq 0,\ mx = my}{x = y}.$$

A similar inference rule could be stated for addition

$$\frac{m + x = m + y}{x = y}.$$

There is no doubt that such inference rules are useful in algebra, and we shall sometimes use them without comment when working with interpretations of the abstract system, but never in the system itself.

6.8 SOLUTION OF A LINEAR EQUATION

A typical statement of an elementary algebra exercise might be

Solve the equation: $2x + 5 = 7$.

There are at least two distinct ways of regarding the problem. One may interpret the injunction "Solve the equation" as meaning transform the given equation by means of certain algebraic rules into an equation of the form "$x =$ a number". The problem is then successfully dealt with if the required activity is carried out to produce "$x = 1$". Alternatively, one may interpret the problem to be: Find all solutions to the equation, which means, find all values for x that satisfy the equation. Then the problem is successfully dealt with if one discovers that 1 is the only such value.

Unfortunately, the words "solve" and "solution" are often used in both senses, and are thus ambiguous. The ambiguity is quite unnecessary. If the first interpretation is desired, the problem could be stated

> Find an equation equivalent to "$2x + 5 = 7$" in the form "x = a constant".

Equivalence of equations would, of course, have to be understood. If the second interpretation is desired, the problem could be stated

> Find all solutions (or roots) of "$2x + 5 = 7$".

"Solution" or "root" would have to be understood as a value for x that satisfies the equation, i.e., yields a true substitution instance for the equation.

In practice, one usually finds the root of an equation by first reducing the equation to an equivalent equation of the form "$x = a$", then clearly "a" is the only root. For the particular equation "$2x + 5 = 7$" we proceed as follows:

1. $2x + 5 = 7$	1. Assumption
2. $2x + 5 = 7 \leftrightarrow 2x = 7 + (-5)$	2. IU, L_3
3. $2x = 7 + (-5)$	3. Mod pon [1,2]
4. $2 \neq 0$	4. Property of the interpretation, i.e., of the real numbers
5. $2 \neq 0 \rightarrow [2x = 7 + (-5) \leftrightarrow x = (7 + (-5))2^{-1}]$	5. IU, L_4
6. $2x = 7 + (-5) \leftrightarrow x = (7 + (-5))2^{-1}$	6. Mod pon [4,5]
7. $x = (7 + (-5))2^{-1}$	7. Mod pon [3,6]

NOTE: If we want to make use of additional properties of real numbers, we may write the equation of Step 7: $x = 1$. So

8. $2x + 5 = 7 \rightarrow x = 1$	8. Ded prin
9. $(x)[2x + 5 = 7 \rightarrow x = 1]$	9. PGU

It is well to pause here to see exactly what has been proved. The foregoing proof demonstrates

$$(x)[2x + 5 = 7 \rightarrow x = 1] \qquad (6.15)$$

The statement (6.15) means that every true substitution instance of "$2x + 5 = 7$" is a true substitution instance of "$x = 1$". While it is perfectly clear that substitution of "1" for "x" yields a true substitution instance of "$x = 1$", we have *not* proved that this will yield a true substitution for "$2x + 5 = 7$". In other words, we have not proved that 1 is a root of the original equation. In order to prove that 1 is a root of "$2x + 5 = 7$" we must prove

$$2 \cdot 1 + 5 = 7. \tag{6.16}$$

There are two ways of accomplishing the proof that 1 is a root. The easiest is to prove (6.16) directly from the addition and multiplication tables for real numbers. Another way is to prove

$$(x)[x = 1 \rightarrow 2x + 5 = 7]. \tag{6.17}$$

We leave the proof as an exercise. After (6.17) is proved, we use IU to get

$$1 = 1 \rightarrow 2 \cdot 1 + 5 = 7,$$

and since "$1 = 1$" results by IU from Axiom R, we have (6.16) by modus ponens.

It is important to observe that while (6.15) does not mean that 1 is a root of "$2x + 5 = 7$", it does mean that any root of "$2x + 5 = 7$" is a root of "$x = 1$", and since 1 is the only root of "$x = 1$", we have that 1 is the only possibility as a root for "$2x + 5 = 7$". Hence (6.15) and (6.16) together establish that 1 is a root, and the only root, of "$2x + 5 = 7$".

Thus, we see that when the elementary algebra student starts with the equation "$2x + 5 = 7$" and arrives by algebraic manipulations at "$x = 1$", he has not found a root of the equation yet. The subsequent checking by substitution of "1" into the equation is a logical necessity, and not just for the purpose of guarding against error. Of course, for simple linear equations such as this, when one is able to infer "$x = r$", it always turns out that r is indeed a root; so it is hard at this stage to bring home to the student the logical necessity for checking. It becomes easier when one has reached more sophisticated equations, where it is possible to infer "$x = r$" but r is not a root of the original equation.

It is worth noting that in the derivation, except in Steps 8 and 9, no properties of the numbers 2, 5, 7 were used that are not properties following from the uninterpreted field axioms. This suggests that it should be possible to prove as a theorem of our abstract system

$$T_6\text{: } (a,b,c,x)\{a \neq 0 \rightarrow [ax + b = c \leftrightarrow x = (c + (-b))a^{-1}]\}.$$

The proof is left as an exercise. It turns out, as before, that $[c + (-b)]a^{-1}$ is a root, and the only root, of "$ax + b = c$".

Since T_6 is a theorem of the abstract system, it can be used to solve equations in any valid interpretation. For example: Find roots of the following equation in the field F[mod 7],

$$4x + 3 = 2. \tag{6.18}$$

Applying T_6, by IU,

$$4x + 3 = 2 \leftrightarrow x = [2 + (-3)]4^{-1}.$$

From the properties of F[mod 7],

$$(-3) = 4,\ 4^{-1} = 2;$$

so

$$[2 + (-3)]4^{-1} = [2 + 4]2 = 6 \cdot 2 = 5,$$

and the required root is 5.

Normally one does not try to remember such theorems as T_6 in an algebra, but solves simple equations like (6.18) by a combination of applications of axioms and cancellation theorems used in abbreviated form. Thus (6.18) could be solved by first multiplying both members by 2 to obtain

$$1 \cdot x + 6 = 4 \tag{6.19}$$

Then add 1 to both members of (6.19) to get

$$x + 0 = 5.$$

EXERCISES

Write out proofs for each of the following lemmas, theorems, or corollaries:

1. L_0, page 198
2. L_1, page 198
3. L_2, page 199
4. T_1, page 199
5. Corollary to T_1, page 199
6. T_2, page 199
7. Corollary to T_2, page 199
8. T_3, page 199
9. Corollary to T_3, page 199
10. Corollary to L_3, page 199
11. L_4, page 199
12. Corollary to L_4, page 199
13. (6.17), page 207
14. T_6, page 207

Find the solutions of the following equations in a form analogous to that used on page 206:

15. $3x + (-5) = 7$

16. $4x + 3 = x + 15$

17. In F[mod 7], $3x + 2 = 4$

18. In F[mod 7], $5x + (-3) = 2x + 6$

19. Prove:

a. $1^{-1} = 1$

b. $x \neq 0 \rightarrow [xy = x \rightarrow y = 1]$

c. $-0 = 0$

d. $(x,y)[x + y = x \rightarrow y = 0]$

e. $(x)[x \neq 0 \rightarrow x^{-1} \neq 0]$

f. $(x)[(x^{-1})^{-1} = x]$

g. $(x,y)[xy \neq 0 \rightarrow (xy)^{-1} = x^{-1}y^{-1}]$

20. Prove: $(x,y)\{(x \neq 0)(y \neq 0) \rightarrow x^{-1} + y^{-1} = (x + y)(xy)^{-1}\}$

6.9 FURTHER THEOREMS IN THE ABSTRACT FIELD SYSTEM

T_7: $(x)[x \cdot 0 = 0]$

1. $1 + 0 = 1$	1. IU, A_4
2. $1 + 0 = 1 \rightarrow x(1 + 0) = x \cdot 1$	2. IU, T_2 Cor.
3. $x(1 + 0) = x \cdot 1$	3. Mod pon [1,2]
4. $x(1 + 0) = x \cdot 1 + x \cdot 0$	4. IU, D_1
5. $x \cdot 1 + x \cdot 0 = x \cdot 1$	5. IS [4,3]
6. $x \cdot 1 = x$	6. IU, M_4
7. $x + x \cdot 0 = x$	7. IS [6,5]
8. $x + x \cdot 0 = x \rightarrow x \cdot 0 = 0$	8. IU, Exercise 19*d*, Sec. 6.8
9. $x \cdot 0 = 0$	9. Mod pon [7,8]
10. $(x)[x \cdot 0 = 0]$	10. PGU

Cor: $(x)[0 \cdot x = 0]$

The next theorem is the basis for solving quadratic equations by the method of factoring.

T_8: $(x,y)[(xy = 0) \rightarrow (x = 0) \vee (y = 0)]$

1. $xy = 0$	1. Assumption
2. $x \neq 0$	2. Assumption
3. $x \neq 0 \rightarrow [xy = 0 \rightarrow y = 0 \cdot x^{-1}]$	3. IU, L_4
4. $xy = 0 \rightarrow y = 0 \cdot x^{-1}$	4. Mod pon [2,3]
5. $y = 0 \cdot x^{-1}$	5. Mod pon [1,4]
6. $0 \cdot x^{-1} = 0$	6. IU, T_7 Cor.

7. $y = 0$	7. IS [6,5]
8. $x \neq 0 \rightarrow y = 0$	8. Ded prin
9. $(x = 0) \vee (y = 0)$	9. Subst. VSF: $\{(\sim P \rightarrow Q) \leftrightarrow (P \vee Q)\}$
10. $(xy = 0) \rightarrow (x = 0) \vee (y = 0)$	10. Ded prin [1,9]
11. T_8	11. PGU

6.10 SOLUTION OF A QUADRATIC EQUATION

As an application of T_8 in the field of real numbers, consider the problem of finding the roots of

$$x^2 - 3x + 2 = 0 \tag{6.20}$$

We shall proceed informally. From the identity,

$$(x)[x^2 - 3x + 2 = (x - 1)(x - 2)],$$

by IU and IS

$$(x - 1)(x - 2) = 0.$$

Then by IU from T_8,

$$[(x - 1)(x - 2) = 0] \rightarrow [(x - 1 = 0) \vee (x - 2 = 0)].$$

So, by modus ponens,

$$(x - 1 = 0) \vee (x - 2 = 0);$$

and finally,

$$(x = 1) \vee (x = 2). \tag{6.21}$$

So, by the deduction principle and PGU,

$$(x)\{[x^2 - 3x + 2 = 0] \rightarrow [(x = 1) \vee (x = 2)]\}. \tag{6.22}$$

It is easy to see that 1 and 2 are numbers yielding true substitution instances of (6.21). That is,

$$(1 = 1) \vee (1 = 2)$$

and

$$(2 = 1) \vee (2 = 2)$$

are both true. Clearly 1 and 2 are the only real numbers yielding true substitution instances of (6.21). Then from (6.22) it follows that there are no roots of (6.20) different from 1 and 2.[1] Direct substitution in (6.20) will verify that the roots are indeed 1 and 2.

[1] Perhaps it is easier to understand the uniqueness of roots by considering a statement equivalent to (6.22): $(x)\{(x \neq 1)(x \neq 2) \rightarrow x^2 - 3x + 2 \neq 0\}$.

Certain errors of language and logic are frequently encountered in the treatment of the foregoing technique. It is quite common to find the solution to (6.20) expressed:

(6.23) The roots of $x^2 - 3x + 2 = 0$ are $x = 1$ and $x = 2$.

In the first place, a root of an equation is a number and is not another equation or statement function; hence, while the roots can be 1 and 2, they certainly cannot be $x = 1$ and $x = 2$. Second, the statement "$x = 1$ and $x = 2$" is surely false.

Proof:

1. $(x = 1)(x = 2)$	1. Assumption
2. $x = 1$	2. Simp, Step 1
3. $x = 2$	3. Simp, Step 1
4. $1 = 2$	4. IS
5. $1 \neq 2$	5. Property of the integers
6. $(1 = 2)(1 \neq 2)$	6. Conj inf

Of course, "$x = 1$ and $x = 2$" is intended to convey the information that 1 and 2 are values of x that satisfy $x^2 - 3x + 2 = 0$. Since it is just as easy to convey the same information by a correct statement, statements such as (6.23) should be avoided.

Since T_8 is a theorem in the abstract system, it is a true statement about any field, and can be used to solve suitable quadratic equations in, for example, F[mod 7]. Consider the solution of

(6.24) $$x^2 + 4x + 2 = 0$$

in F[mod 7]. Then, informally,

$$(x + 6)(x + 5) = x^2 + (6 + 5)x + 2 = x^2 + 4x + 2.$$

So

$$(x + 6)(x + 5) = 0,$$
$$(x + 6 = 0) \vee (x + 5 = 0),$$
$$(x = 1) \vee (x = 2).$$

Then, since

$$1^2 + 4 \cdot 1 + 2 = 0 \text{ and } 2^2 + 4.2 + 2 = 0,$$

the roots of (6.24) are 1 and 2.

The set of numbers $\{0,1,2,3,4,5\}$ with addition and multiplication modulo 6 is not a valid interpretation of the abstract field system (for example, M_5 does not hold) and T_8 is not a true statement about this set and its operations, as is shown by the counter example

$$2 \cdot 4 = 0.$$

Thus T_8 is not available for solving quadratic equations in the modulo 6 system. For example, suppose we attempt to solve the equation

$$x^2 + 5x = 0, \tag{6.25}$$

in the system modulo 6. It is easy to show the identity

$$(x)\{x^2 + 5x = (x + 2)(x + 3)\},$$

and obtain by substitution

$$(x + 2)(x + 3) = 0. \tag{6.26}$$

Now, lacking T_8 for this system, we cannot deduce from (6.26)

$$(x = 4) \vee (x = 3), \tag{6.27}$$

and in fact neither (6.25) nor (6.26) implies (6.27), since both 0 and 1 yield true substitution instances of (6.25) and (6.26) but not of (6.27). In fact, the roots of (6.25) are 0, 1, 3, 4, but only two of them satisfy (6.27).

It is also worth noting that the quadratic polynomial "$x^2 + 5x$" can be factored into two linear factors different from those of (6.26), i.e.,

$$x^2 + 5x = x(x + 5).$$

Thus there is a second-degree polynomial in the system modulo 6 that has more than two distinct roots and more than one factorization into linear factors. Such a situation is not possible in a field.

The foregoing examples suggest that many of the basic theorems of algebra may depend only on field properties. Furthermore, however obvious T_8 may seem, it expresses a basic and important property of elementary algebra.

EXERCISES

1. Prove the corollary to T_7, page 209.

2. Fill in the details of the solution of the quadratic equation given on page 210.

Find all the roots of the following equations:

3. $x^2 + 4x + (-5) = 0$ in F[mod 7]

4. $x^2 + 5x = 0$ in F[mod 7]

5. $x^2 + 4x = 0$ in the system modulo 6

6. $x^2 + 3x + 2 = 0$ in the system modulo 6

7. $x^2 + 2x + (-1) = 0$ in F[mod 3]

8. $2x^2 + 3x + 4 = 0$ in F[mod 5]

9. Prove: $(x,y)[xy \neq 0 \rightarrow (x \neq 0)(y \neq 0)]$

6.11 SUBTRACTION AND DIVISION

The inverse operations of *subtraction* and *division* may be introduced into the abstract system by definition.

Definition of Subtraction:

$$(x,y)[x - y = x + (-y)]$$

Definition of Division:

$$(x,y)[y \neq 0 \rightarrow (x \div y = xy^{-1})]$$

In the definition of subtraction, the symbol "$-$" enters ambiguously. In "$x - y$" the symbol "$-$" is used to denote the binary operation of *subtraction*, but in "$(-y)$" the same symbol is used to denote the singulary operation of *negation*. While the two operations are closely related, they are not the same, and strictly should be denoted by different symbols. It is customary in algebra, however, to use the same symbol for both operations and to depend on context to determine which operation is meant. The rule to follow is:

a. When "$-$" occurs between two terms, the binary operation of subtraction is indicated.
b. When "$-$" occurs before a single term, the singulary operation of negation is indicated.

Such expressions as

$$y + -x, \qquad 3 - -y, \qquad y - , \qquad - -2,$$

are not properly formed and will not be used as abbreviations. On the other hand, the expression

$$x - y - z - w$$

is customarily used as an abbreviation for

$$[(x - y) - z] - w$$

and

(6.28) $$10 - 2 + 3 - 4 + 2$$

is an abbreviation for

(6.29) $$\{[(10 - 2) + 3] - 4\} + 2.$$

Such mixed expressions are not ambiguous since it is understood that association is to the left. So, for instance, (6.28) is our abbreviation for

(6.29), while the following are not:

$$[10 - (2 + 3)] - (4 + 2), \qquad 10 - \{2 + [3 - (4 + 2)]\}.$$

All the above are familiar conventions of algebra. They are practical abbreviations and not hard to learn. It seems unfair, however, to expect the algebra student to learn them without some explicit exposition by the teacher. Many of the usual errors committed by algebra students result from confusion about these conventions.

With the foregoing definitions, it is easy to deduce properties of subtraction and division. We list some without proof:

L_5: $(x)[0 - x = (-x)]$
L_6: $(x)[x \neq 0 \rightarrow (1 \div x = x^{-1})]$
L_7: $(x,y,z)[x + y = z \leftrightarrow x = z - y]$
L_8: $(x,y,z)[y \neq 0 \rightarrow (xy = z \leftrightarrow x = z \div y)]$
L_9: $(x,y,z)[x(y - z) = xy - xz]$

6.12 SOLUTION OF SIMULTANEOUS LINEAR EQUATIONS

The equation in two variables

$$x + y = 10 \tag{6.30}$$

can be thought of as an equation in the field of real numbers if the universe of individuals is limited to the real numbers. Various substitution instances of (6.30) are

(6.31)

a. $2 + 11 = 10$
b. $7 + 3 = 10$
c. $3 + 7 = 10$
d. $1 + 1 = 10$
e. $12 + (-2) = 10$

The false substitution instance (6.31*a*) is obtained by replacing "x" by "2" and "y" by "11" in (6.30). It is convenient to say that substitution of "(2,11)" in (6.30) yields (6.31*a*), where we agree that the first member of the ordered pair is substituted for "x" and the second member of the ordered pair is substituted for "y". Similarly, (7,3) yields (6.31*b*), (3,7) yields (6.31*c*), and so on.

We call any ordered pair a solution of (6.30) if it yields a true substitution instance of (6.30).

The injunction: Solve the system:

(6.32)

$$x + y = 10$$
$$x - y = 6$$

has the same ambiguity of meaning as for the case of a single equation in one variable. We shall interpret the injunction to mean

> Find all ordered pairs of numbers yielding true substitution instances for "$(x + y = 10)$ and $(x - y = 6)$".

First we prove

(6.33) $\quad (x,y)[(x + y = 10)(x - y = 6) \rightarrow (x = 8)(y = 2)].$

The proof will be abbreviated by multiple use of IS and by suppressing major premises of modus ponens. The multiplication and addition tables for integer numbers are assumed known.

	1. $(x + y = 10)(x - y = 6)$	1. Assumption
	2. $x + y = 10$	2. Simp Step 1
	3. $x - y = 6$	3. Simp Step 1
	4. $x = 6 + y$	4. IU, L_7; mod pon [3]
	5. $(6 + y) + y = 10$	5. IS [4,2]
	6. $(6 + y) + y = 6 + (y + y)$	6. IU, A_2
or:	6′. $(6 + y) + y = 6 + 2y$	6′. Property of real numbers; i.e., $1 + 1 = 2$
	7. $6 + 2y = 2y + 6$	7. IU, A_3
	8. $2y + 6 = 10$	8. IS [7,6′,5]
	9. $2y = 10 + (-6)$	9. IU, L_3; mod pon [8]
or:	9′. $2y = 4$	9′. Property of real numbers
	10. $2 \neq 0$	10. Property of real numbers
	11. $y = 2^{-1} \cdot 4$	11. IU, L_4 Cor.; mod pon [10,9′]
or:	11′. $y = 2$	11′. Property of real numbers
	12. $x + 2 = 10$	12. IS [11′,2]
	13. $x = 10 + (-2)$	13. IU, L_3; mod pon [12]
or:	13′. $x = 8$	13′. Properties of real numbers
	14. $(x = 8)(y = 2)$	14. Conj inf [11′,13′]
	15. $(x + y = 10)(x - y = 6) \rightarrow (x = 8)(y = 2)$	15. Ded prin
	16. $(x,y)[(x + y = 10)(x - y = 6) \rightarrow (x = 8)(y = 2)]$	16. PGU

It is clear that (8,2) is the only ordered pair that yields a true substitution instance of

$$(x = 8)(y = 2).$$

Hence, if the original pair of equations has any solution, it is (8,2). Verification is by direct substitution:

$$(8 + 2 = 10)(8 - 2 = 6).$$

The proof of (6.33) is long even with the abbreviations. The object in presenting it is to show how the familiar procedure for solving the system (6.32) depends upon abstract field properties, and incidently to point out those spots where additional special properties of the real numbers are needed.

If one merely wants to find a solution for the system (6.32), the best way is to guess and check the guess; or if some sort of derivation is required, one would omit any mention of associative and commutative properties and perhaps write down Steps 2, 3, 4, 5, 8, 9′, 11′, 12, 13′, and then check the solution suggested by 11′ and 13′.

The proof of (6.33) suggests that in solving a system of simultaneous equations, one is really starting from a conjunction of the two equations. One derives a simple conjunction of the form "$(x = a)(x = b)$". This establishes uniqueness of the solution. Then one checks to establish existence of (a,b) as the solution.

There are, of course, systems of linear equations having no solution. The proof, in such a case, is quite straightforward. Consider the system

$$\begin{aligned} x + y &= 10 \\ x + y &= 6 \end{aligned} \tag{6.34}$$

We start by assuming the conjunction

$$(x + y = 10)(x + y = 6), \tag{6.35}$$

and proceed informally to get the separate equations by simplification. Then

$$x = 6 - y \tag{6.36}$$

and by substitution of (6.36) in the first equation of (6.34)

$$(6 - y) + y = 10$$

which yields

$$6 = 10.$$

But $6 \neq 10$, so we have a contradiction, and the logical negation of (6.35) is established. Then by PGU

$$(x,y)\sim[(x + y = 10)(x + y = 6)],$$

or by conversion of quantifiers

$$\sim(\exists x,y)[(x + y = 10)(x + y = 6)].$$

The proof is quite straightforward, and does not require geometric arguments leaning on parallelism. We do not want to discount the value of geometric interpretations in teaching algebra, but merely want to point out that the inconsistency of the system (6.34) is a fact of the real-number field, quite independent of geometric notions.

The procedures for solving systems of equations in the real-number field should be effective in any field, taking into account the addition and multiplication tables of the field. For example, consider the following system in $F[\text{mod } 5]$:

$$\begin{aligned} 3x + 4y &= 2 \\ 2x + 3y &= 1 \end{aligned} \tag{6.37}$$

We solve the system (6.37) informally, and start by multiplying both members of the second equation by 3 to obtain

$$x + 4y = 3.$$

Then, by L_7,

$$x = 3 - 4y \tag{6.38}$$

and, by substitution in the first equation of (6.37),

$$3(3 - 4y) + 4y = 2;$$

or

$$4 - 2y + 4y = 2,$$

or

$$4 + 2y = 2.$$

Then, adding 1 to both members, and then multiplying both members by 3,

$$y = 4.$$

Substitution in (6.38) yields

$$x = 2.$$

Thus the system (6.37) implies the system

$$\begin{aligned} x &= 2 \\ y &= 4. \end{aligned} \tag{6.39}$$

The only solution of (6.39) is clearly (2,4), and direct substitution in (6.37) shows that (2,4) is a solution, and hence the only solution of (6.37).

EXERCISES

Write out a proof analogous to (6.33) for each of the following systems of equations:

1. $\begin{aligned} x + y &= 7 \\ 2x - y &= 3 \end{aligned}$

2. $\begin{aligned} 3x + y &= 6 \\ 2x - 3y &= 4 \end{aligned}$ in $F[\text{mod } 7]$

Solve each of the following systems of equations in the indicated field:

3. $2x + y = 4$, $x - y = 0$ in $F[\text{mod } 5]$

4. $3x + 6y = 1$, $2x - 4y = 5$ in $F[\text{mod } 7]$

5. $2x + y = 1$, $x + y = 0$ in $F[\text{mod } 3]$

6. $4x - y = 2$, $3x + y = 4$ in $F[\text{mod } 5]$

7. Show that if *division* is associative in a field, then

$$(z)[z^2 = 1]$$

can be proved. Then find a field having this property.

8. Prove: $(x,y,z)\{(x - y) + z = x + (z - y)\}$.

9. Prove: In the field of real numbers

$$\sim(\exists x,y)[(x + y = 10)(x + y = 6)].$$

6.13 NEGATION VS. NEGATIVE NUMBER

One of the mysteries of elementary algebra is this: Why is it that $(-2)(-2) = +4$? Removing the mystery from this question poses problems to the teacher because the student normally builds his conception of numbers by abstracting from many different interpretations of numbers. Unfortunately, such interpretations of positive and negative numbers as credits and debits, or temperature, etc., that work so well in teaching addition and subtraction of signed numbers are not so satisfactory for dealing with multiplication of signed numbers, particularly for the case of $(-2)(-2)$.

In fact, the answer is to be found in formal considerations. Essentially, $(-2)(-2) = +4$ because it is defined that way. Still, one can ask why is the product defined this way rather than as -4, or 7, or $\frac{1}{2}$ or $-\pi$? To answer this, we must consider how the integers are themselves defined.

The natural numbers 1, 2, 3, . . . with ordinary addition and multiplication have all the properties expressed in the field axioms A_1, A_2, A_3, M_1, M_2, M_3, M_4, D_1. Subtraction and division can be defined for some pairs of natural numbers, but not for others. One can now try to extend the natural numbers to a larger system in which subtraction is always defined by creating some new numbers to serve as differences such as $1 - 3$, $3 - 10$, etc., which are not defined for natural numbers. The requirement that subtraction be always possible is equivalent to the requirement that A_4 and A_5 hold for the system. Of course, one wants the new system to satisfy all the axioms satisfied by the natural numbers. But with all these axioms satisfied, it is possible to prove

(6.40)

$$a.\ (x,y)[(-x)y = -(xy)]$$
$$b.\ (x,y)[(-x)(-y) = xy]$$

where the "$-$" denotes the negation operation as defined for fields.

The procedure for extending the natural numbers might then be:

1. Assume an integer denoted by "0" satisfying A_5 and D_2.
2. Call the natural numbers *positive integers*, and assume for each positive integer x a corresponding *negative integer*, to be denoted (for the moment) by "$\bar{x}$", with the property that $\bar{x} = -x$. [Remember, "$-$" denotes the negation operation.][1]

If addition and multiplication of the signed numbers are now defined in the usual way, the resulting system will be found to satisfy all of the required axioms. In particular, one easily proves, $-\bar{x} = x$.

It is worth noting that the property that the product of two negative integers is positive is in a sense a special case of (6.40*b*), which is a statement about all integers. By definition of $\overline{2}$ and $\overline{3}$, we have $\overline{2} = (-2)$ and $\overline{3} = (-3)$; hence from (6.40*b*) we obtain: $\overline{2} \cdot \overline{3} = 2 \cdot 3$. But from (6.40*b*), one also can obtain: $(-\overline{2})(-3) = \overline{2} \cdot 3$ and $(-\overline{2})(-\overline{3}) = \overline{2} \cdot \overline{3}$. In fact, (6.40*a*) and (6.40*b*) can be proved as consequences of A_1 to A_5, M_1 to M_4 and D_1, D_2 without any notion of positive and negative numbers. As we are about to show, (6.40*a*) and (6.40*b*) are theorems of the abstract field system, and hence express properties of all fields. Yet in the field F[mod 5] the notion of positive or negative numbers is meaningless (see Sec. 6.14).

We now prove (6.40*a*) and (6.40*b*) as Theorems 9 and 11 for abstract fields.

T_9: $(x,y)[(-x)y = -(xy)]$

Proof: (Abbreviation through multiple use of IS)

1. $x + (-x) = 0$	1. IU, L_1
2. $x + (-x) = 0 \rightarrow [x + (-x)]y = 0 \cdot y$	2. IU, T_2
3. $0 \cdot y = 0$	3. IU, T_8 Cor.
4. $[x + (-x)]y = xy + (-x)y$	4. IU, T_4
5. $xy + (-x)y = 0$	5. IS [3,4 in 2]
6. $(xy + (-x)y = 0) \rightarrow ((-x)y = -(xy))$	6. IU, Def of additive inverse
7. $(-x)y = -(xy)$	7. Mod pon [5,6]
8. $(x,y)[(-x)y = -(xy)]$	8. PGU

Cor: $(x,y)[x(-y) = -(xy)]$

[1] A more elegant method for extending the natural numbers is to deal with a set whose elements are ordered pairs of natural numbers. A short account of this method can be found in: *The Twenty-third Yearbook of the National Council of Teachers of Mathematics*, "Insights Into Modern Mathematics," chap. II, Washington, D.C.: National Council of Teachers of Mathematics, 1957.

T_{10}: $(x)[x = -(-x)]$

1. $x + (-x) = 0$	1. IU, L_1
2. $x + (-x) = (-x) + x$	2. IU, A_3
3. $(-x) + x = 0$	3. IS [2,1]
4. $(-x) + x = 0 \rightarrow x = -(-x)$	4. IU, Def of additive inverse
5. $x = -(-x)$	5. Mod pon [3,4]
6. $(x)[x = -(-x)]$	6. PGU

T_{11}: $(x,y)[(-x)(-y) = xy]$

1. $(-x)(-y) = -(x(-y))$	1. IU, T_9
2. $x(-y) = -(xy)$	2. IU, T_9, Cor.
3. $(-x)(-y) = -(-(xy))$	3. IS [2,1]
4. $-(-(xy)) = xy$	4. IU, T_{10}
5. $(-x)(-y) = xy$	5. IS [4,3]
6. $(x,y)[(-x)(-y) = xy]$	6. PGU

6.14 ORDERED FIELDS

The field of real numbers has all the properties expressed in the axioms and theorems of the abstract field system, but has other properties that are not deducible from these field axioms. Among these are:

a. The field has infinitely many elements.
b. There exist real numbers that are not the sum of two squares, e.g., -4.
c. The field contains a proper subfield, e.g., the field of rational numbers.
d. The real numbers are ordered by the relation "$<$" (less than).

Since the field F[mod 5] has none of these properties, it follows that none of the properties *a*, *b*, *c*, *d* are deducible from the field axioms.

It is not hard to verify that F[mod 5] does not have properties (*a*), (*b*), (*c*). Indeed, it is obvious in the case of (*a*). As for (*b*), in F[mod 5] one can write

$$\begin{aligned} 0 &= 0^2 + 0^2 \\ 1 &= 0^2 + 1^2 \\ 2 &= 1^2 + 4^2 \\ 3 &= 2^2 + 2^2 \\ 4 &= 0^2 + 3^2 \end{aligned}$$

Hence, every element of the field is the sum of two squares.

In order to check property (*c*), we must first define *subfield*. A subfield H of a field F is a subset of F that forms a field under the operations of F. H is called *proper* if H is not the same as F. Now F[mod 5] contains no proper subfield, because any subfield would have to contain the unit elements 0 and 1, and hence would also (by A_1) have to contain $1 + 1$, and hence $2 + 1$, and $3 + 1$. That is, any subfield of F[mod 5] contains all the elements of F[mod 5], and is thus not a proper subfield.

To show that F[mod 5] cannot be ordered, consider how 1 and 0 would be ordered, if this were possible. Since $1 \neq 0$, we should expect either that $0 < 1$ or that $1 < 0$, but not both. Suppose $0 < 1$. Then adding to both members (mod 5 of course) should preserve the inequality in the same direction. Repeating this procedure five times yields the inequalities

$$0 < 1,\ 1 < 2,\ 2 < 3,\ 3 < 4,\ 4 < 0.$$

The last four of these inequalities together imply that $1 < 0$. So we have

$$0 < 1 \rightarrow 1 < 0.$$

Similarly it can be shown that

$$1 < 0 \rightarrow 0 < 1;$$

so finally

$$1 < 0 \leftrightarrow 0 < 1.$$

Since this result does not accord with what we expect of an order relation, we conclude that F[mod 5] cannot be ordered.

The foregoing discussion made tacit use of several properties of an order relation. We now state the properties formally as axioms. These axioms, added to the axioms for an abstract field, yield the axioms for an *abstract ordered field.* Any interpretation of the resulting abstract system is called an *ordered field.* The real numbers form such an ordered field, but as we have seen, F[mod 5] does not. Indeed, for all positive primes p, F[mod p] is not ordered.

The Order Axioms ["$x < y$" is read "x is less than y"]:

O_1: $(x,y)[(x < y)\vee(x = y)\vee(y < x)]$
O_2: $(x,y)[(x < y) \rightarrow \sim(y < x)]$
O_3: $(x,y,z)[(x < y)(y < z) \rightarrow (x < z)]$
O_4: $(x,y,z)[(x < y) \rightarrow (x + z < y + z)]$
O_5: $(x,y,z)\{(0 < z) \rightarrow [(x < y) \rightarrow (zx < zy)]\}$

The axioms O_1 and O_2 together express the *trichotomous* property of "$<$", for from them it will be possible to prove (see T_{18}) that the conditions "$x < y$", "$x = y$", "$y < x$" are mutually exclusive.

O_3 expresses the *transitive* property of "$<$"; O_4 and O_5 connect the order relation with the binary operations and permit the usual manipulations of inequalities.

The order axioms express familiar properties of the order relation in the field of real numbers. We proceed to prove some theorems of the abstract ordered field system. In particular, we shall show that it makes sense to talk about *positive* and *negative* elements in any ordered field. "x is positive" will be interpreted to mean "$0 < x$", and "x is negative" to mean "$x < 0$". We do not introduce any new symbol to indicate a positive or a negative element; "$+$" always denotes the field operation of addition, and "$-$" either the binary operation of subtraction, or the singulary operation of negation, depending on context.

T_{12}: $(x)[0 < x \rightarrow -x < 0]$

1. $0 < x \rightarrow 0 + (-x) < x + (-x)$	1. IU, O_4
2. $0 + (-x) = (-x)$	2. IU, A_4, and commutativity
3. $x + (-x) = 0$	3. IU, L_1
4. $0 < x \rightarrow (-x) < 0$	4. IS [2,3 in 1]
5. T_{12}	5. PGU

Cor: $(x)[x < 0 \rightarrow 0 < -x]$

One can interpret T_{12} to mean that if x is positive, then $-x$ is negative; and the corollary to mean that if x is negative, then $-x$ is positive.

By repeated applications of O_4, it can be proved that

T_{13}: $(x,y)[x < y \leftrightarrow -y < -x]$

The proof is left as an exercise.

The next lemmas and Theorem 14 establish that in any ordered field, every square is positive except 0^2.

L_{10}: $(x)[0 < x \rightarrow -x < x]$

Proof is left as an exercise.

L_{11}: $(x)[x \neq 0 \rightarrow -x^2 < x^2]$

1. $x \neq 0$	1. Assumption
2. $(x < 0) \vee (x = 0) \vee (0 < x)$	2. IU, O_1
3. $x \neq 0 \rightarrow (x < 0) \vee (0 < x)$	3. Subst. VSF
4. $(x < 0) \vee (0 < x)$	4. Mod pon [1,3]
5. Case I: $0 < x$	5. Assumption
6. $0 < x \rightarrow [-x < x \rightarrow x(-x) < x^2]$	6. IU, O_5
7. $-x < x \rightarrow x(-x) < x^2$	7. Mod pon [5,6]
8. $0 < x \rightarrow -x < x$	8. IU, L_{10}

9. $-x < x$	9. Mod pon [5,8]
10. $x(-x) < x^2$	10. Mod pon [9,7]
11. $x(-x) = -x^2$	11. IU, T_9 Cor.
12. $-x^2 < x^2$	12. IS [11,10]
13. $0 < x \rightarrow -x^2 < x^2$	13. Ded prin [assump. Case I disch'd]
14. Case II: $x < 0$	14. Assumption
15. $x < 0 \rightarrow 0 < -x$	15. IU, T_{12} Cor.
16. $0 < -x$	16. Mod pon [14,15]
17. $0 < -x \rightarrow [x < -x \rightarrow (-x)x < (-x)(-x)]$	17. IU, O_5
18. $x < -x \rightarrow (-x)x < (-x)(-x)$	18. Mod pon [16,17]
19. $0 < -x \rightarrow -(-x) < -x$	19. IU, L_{10}
20. $-(-x) < -x$	20. Mod pon [16,19]
21. $x = -(-x)$	21. IU, T_{10}
22. $x < -x$	22. IS [21,20]
23. $(-x)x < (-x)(-x)$	23. Mod pon [22,18]
24. $(-x)x = -x^2$	24. IU, T_9
25. $(-x)(-x) = x^2$	25. IU, T_{11}
26. $-x^2 < x^2$	26. IS [24,25 in 23]
27. $x < 0 \rightarrow -x^2 < x^2$	27. Ded prin [assump. Case II disch'd]
28. $(x < 0) \vee (0 < x) \rightarrow -x^2 < x^2$	28. Inf by cases [13,27]
29. $-x^2 < x^2$	29. Mod pon [4,28]
30. $x \neq 0 \rightarrow -x^2 < x^2$	30. Ded prin [1,29]
31. L_{11}	31. PGU

L_{12}: $(x)[x^2 = 0 \rightarrow x = 0]$

Proof is left as an exercise. (Use T_8.)

T_{14}: $(x)[x \neq 0 \rightarrow 0 < x^2]$

1. $x \neq 0$	1. Assumption
2. $x^2 = 0 \rightarrow x = 0$	2. IU, L_{12}
3. $x^2 \neq 0$	3. Contrap inf [2,1]
4. $(x^2 < 0) \vee (x^2 = 0) \vee (0 < x^2)$	4. IU, O_1
5. $x^2 \neq 0 \rightarrow (x^2 < 0) \vee (0 < x^2)$	5. Subst. VSF
6. $(x^2 < 0) \vee (0 < x^2)$	6. Mod pon [3,5]
7. $\sim(x^2 < 0) \rightarrow 0 < x^2$	7. Subst. VSF for 6
8. $x^2 < 0$	8. Assumption
9. $x \neq 0 \rightarrow -x^2 < x^2$	9. IU, L_{11}
10. $-x^2 < x^2$	10. Mod pon [1,9]
11. $(-x^2 < x^2)(x^2 < 0) \rightarrow -x^2 < 0$	11. IU, O_3

12. $-x^2 < 0$	12. Conj inf [8,10] and mod pon [11]
13. $-x^2 < 0 \rightarrow 0 < -(-x^2)$	13. IU, T_{12} Cor.
14. $0 < -(-x^2)$	14. Mod pon [12,13]
15. $-(-x^2) = x^2$	15. IU, T_{10}
16. $0 < x^2$	16. IS [15,14]
17. $x^2 < 0 \rightarrow 0 < x^2$	17. Ded prin [assump 8 disch'd]
18. $(x^2 < 0) \vee \sim(x^2 < 0) \rightarrow 0 < x^2$	18. Inf by cases [17,7]
19. $(x^2 < 0) \vee \sim(x^2 < 0)$	19. VSF
20. $0 < x^2$	20. Mod pon [19,18]
21. $x \neq 0 \rightarrow 0 < x^2$	21. Ded prin
22. T_{14}	22. PGU

Since $1 = 1^2$, it follows immediately from T_{14} that 1 is positive. From this result and T_{12} it follows that -1 is negative, hence,

T_{15}: $0 < 1$

Cor: $-1 < 0$.

Theorems T_{12}, T_{14}, T_{15} provide another means for proving that F[mod 5] cannot be ordered. For, suppose it is ordered. Then, by T_{15}, $0 < 1$ and, from T_{12}, $4 < 0$ (since $-1 = 4$ in this field). This result conflicts with T_{14}, since $4 = 2^2$.

The same theorems allow us to prove that the field of complex numbers is not an ordered field. Suppose the complex numbers are ordered by a binary relation "$<$" satisfying the order axioms. Now the multiplicative unit for the field is $1 + 0 \cdot i$ or 1. Hence by T_{15}, $0 < 1$ and by the corollary $-1 < 0$. But in this field $i^2 = -1$; hence $i^2 < 0$, and by T_{14}, $i^2 > 0$. From these results a contradiction is easy to deduce, and it follows that the complex numbers cannot be ordered. This is not to say that *no* sort of ordering can be introduced into the complex field, but whatever order relation is established, it will not satisfy all the axioms O_1 to O_5. One might, for instance, define

$$a + bi \circledless c + di \text{ if and only if } a^2 + b^2 < c^2 + d^2.$$

The relation "$\circledless$" has some of the usual order properties, but not O_1, for example.

In an ordered field, the connection between *order* and *difference* is easily established:

T_{16}: $(x,y)[x < y \leftrightarrow 0 < y - x]$

The theorem follows from O_4 and the definition of subtraction.

EXERCISES

1. Prove: T_{12} Cor., page 222.
2. Prove T_{13}, page 222.
3. Prove L_{10}, page 222.
4. Prove L_{12}, page 223.
5. Prove T_{16}, page 224.
6. Show that F[mod 7] cannot be ordered by finding two numbers whose squares add up to -1 and then proving this is a contradiction.
7. Write out an abbreviated proof of T_{14} that could be understood without knowledge of symbolic logic.

Abbreviations involving the symbols "$>$", "$\leq$", "$\geq$" are quite useful:

Statement function	*Abbreviation*	*Reading of abbreviation*
$x < y$	$y > x$	y is greater than x.
$(x < y) \vee (x = y)$	$x \leq y$	x is less than or equal to y.
$(x = y) \vee (x > y)$	$x \geq y$	x is greater than or equal to y.
$(x < y)(y < z)$	$x < y < z$	y is between x and z.
$(x \leq y)(y < z)$	$x \leq y < z$	No customary reading
$(x < y)(y \leq z)$	$x < y \leq z$	No customary reading
$(x \leq y)(y \leq z)$	$x \leq y \leq z$	No customary reading

With these abbreviations, the trichotomy axioms O_1 and O_2 would be written

O_1: $(x,y,z)[(x < y) \vee (x = y) \vee (x > y)]$ or
$(x,y,z)[(x < y) \vee (x \geq y)]$, etc.
O_2: $(x,y)[(x < y) \rightarrow \sim(x > y)]$†

and O_3 becomes

O_3: $(x,y,z)[x < y < z \rightarrow x < z]$.

We shall use these abbreviations in the next theorem, and freely from now on.

T_{17}: $(x,y)[\sim(x < y) \leftrightarrow (x \geq y)]$

1. $(x < y) \vee (x = y) \vee (x > y)$	1. IU, O_1
2. $\sim(x < y) \rightarrow x \geq y$	2. Subst. VSF $\{P \vee Q \leftrightarrow (\sim P \rightarrow Q)\}$
3. $(x = y)(x < y)$	3. Assumption {for indirect proof}
4. $x = y$	4. Simp [3]
5. $x < y$	5. Simp [3]

† A very common abbreviation for "$\sim(x < y)$" is "$x \nless y$".

6. $x < x$	6. IS [4,5]
7. $x < x \rightarrow \sim(x < x)$	7. IU, O_2
8. $\sim(x < x)$	8. Mod pon [6,7]
9. $(x < x)\sim(x < x)$ CONTRA	9. Conj inf [6,8]
10. $\sim[(x = y)(x < y)]$	10. By indirect proof, Steps 3 to 9 [assump 3 disch'd]
11. $x = y \rightarrow \sim(x < y)$	11. Subst. VSF for 10
12. $x > y \rightarrow \sim(x < y)$	12. IU, O_2 {full use of abbreviations}
13. $x \geq y \rightarrow \sim(x < y)$	13. Inf by cases [11,12]
14. $\sim(x < y) \leftrightarrow (x \geq y)$	14. Conj inf [2,13] and def "$\leftrightarrow$"
15. T_{17}	15. PGU

This theorem not only gives a useful form for "$\sim(x < y)$" but the internal indirect proof, Steps 3 to 10, shows that the cases "$x = y$" and "$x < y$" of O_1 are exclusive. Clearly it follows that all the cases of O_1 are mutually exclusive.

We could go on to prove equivalences expressing negations of "$x > y$", "$x \geq y$", "$x \leq y$", but in fact they are all contrapositives or abbreviated forms of T_{17}. Thus, when we need to negate any of the foregoing, we shall refer to T_{17}.

The remaining theorems of this section are stated without proof. They express familiar properties of real numbers that are properties of any ordered field.

T_{18}: $(x,y,z,w)[(x < y)(z < w) \rightarrow x + z < y + w]$
Cor: $(x,y,z,w)[(x \leq y)(z \leq w) \rightarrow x + z \leq y + w]$

T_{19}: $(x,y,z,w)[(x < y)(z > w) \rightarrow x - z < y - w]$
Cor: $(x,y,z,w)[(x \leq y)(z \geq w) \rightarrow x - z \leq y - w]$

Observe that "$\rightarrow$" may not be replaced by "$\leftrightarrow$" in T_{18}, T_{19}.

T_{20}: $(x)[0 < x \leq 1 \rightarrow x \geq x^2]$
T_{21}: $(x)[x \geq 1 \rightarrow x^2 \geq x]$
T_{22}: $(x)[1 \leq x \rightarrow 1 \geq x^{-1}]$

EXERCISES

1. Verify that the relation "$\otimes$" defined for complex numbers, page 224, does not satisfy all the order axioms.

2. Prove T_{18}.

3. Prove T_{18} Cor.

4. Prove T_{19}.

5. Prove T_{19} Cor.

6. Prove T_{20}, page 226.
7. Prove T_{21}, page 226.
8. Prove T_{22}, page 226.
9. Prove: $(x,a,b,c)[(x < a + b)(b < c) \rightarrow x < a + c]$.
10. Prove: $(x,y)\{(x \neq 0)(y \neq 0) \rightarrow [x < y \rightarrow x^{-1} > y^{-1}]\}$.
11. Verify informally that the rational numbers form a proper subfield of the real numbers.

6.15 ABSOLUTE VALUE

The successful algebra student ultimately comes to understand the order relation in the real numbers. He understands that "a" (or even "$+a$") can have negative substitution instances, and that "$-a$" can have positive substitution instances. He is familiar with the interpretation of real numbers as points on a line, with "$a < b$" interpreted as "a to the left of b". He probably has a workable concept of the distinction between *subtraction* and *negation*. These concepts are quite sophisticated compared with those he acquired when first approaching integer operations. For example, the concept of $+2$ and -2 as different aspects of 2 ultimately is replaced by the concept of $+2$ and -2 as integers that are negations of each other.

Generally, the language appropriate to the more sophisticated concepts will be different from the language used in the initial approach. In this initial treatment of integer operations, the teacher wants to build on his students' knowledge of natural number operations and relations. Consequently, operational rules for integers are stated in terms of natural-number operations and the natural-number order relation as far as possible. One of the rules for addition of integers is often condensed into the statement

(6.41) To add two integers of unlike signs, take the difference and attach the sign of the larger.

A little analysis of (6.41) shows that the language is quite poor. Let us apply (6.41) to obtain the sum of the integers $+5$ and -7. To begin with, $+5$ and -7 do not have unlike signs, it is the names "$+5$" and "-7" that have the unlike signs. The connection between the *signs* and the *integers* is just an accident of the method of naming the integers. What are the signs of "$-(-5)$", "$+(5 - 12)$", for example? It appears that (6.41) is a statement concerned with a particular way of representing integers. To go on with the rule, one now has to "take the difference", but clearly not the difference of $+5$ and -7, but of 7 and 5. At this early stage, the student probably doesn't distinguish between "the difference of 7 and 5" and "the difference of 5 and 7", so he gets the difference 2.

The larger of 5 and 7 is 7, so he attaches a " $-$ " to the name "2" to get " -2" and the answer -2. Of course, it is the larger of 5 and 7 that is in question here, and not the larger of $+5$ and -7 which is what (6.41) really says to consider.

One of the troubles with (6.41) is that the antecedent of *difference* and *larger* is not the *two integers*. What is needed is some convenient name for the positive integer (or, if one likes, the natural number) associated with each integer. Thus, in applying rule (6.41) to the sum of $+5$ and -7, one operates with $+5$ and $+7$ (or 5 and 7) which are sometimes called the *numerical values* of $+5$ and -7, respectively.

The term "numerical value" in this context is largely a term confined to elementary or secondary mathematics. It is virtually never used in higher mathematics, where the term "absolute value" is used instead. The absolute value of x is denoted by " $|x|$ " and can be defined for any ordered field as follows:

Definition of absolute value (numerical value):

$$(6.42) \qquad (x)[(x \geq 0 \rightarrow |x| = x)(x < 0 \rightarrow |x| = -x)]$$

Note that " $-$ " indicates the negation operator in (6.42). Statement (6.42) may be paraphrased:

> If x is positive or zero, the absolute value of x is just x, and if x is negative, then the absolute value of x is the negation of x.

Using this notion of absolute value, (6.41) may be restated:

> To form a name for the sum of a positive and a negative integer, find the absolute value of the difference of their absolute values. Then attach a " $-$ " or a " $+$ " sign according as the absolute value of the negative integer is greater than, or less than, the absolute value of the positive integer.

The statement is correct, but a pedagogical horror. An altogether simpler statement is:

> To add a positive and a negative integer, from the integer of larger absolute value subtract the negation of the other integer.

For this rule to be effective, the student must know how to subtract -2 from -5, and -7 from -12, etc., but not, for example, -12 from -7.

However the language problems are solved in the statements of these rules, the rules are intimately concerned with the order relation among positive numbers. Where the language used at this stage is loose, the student is likely to acquire a deep down belief that -7 is greater than $+5$, etc. In some students this belief dies hard.

EXERCISES

1. Write out a set of rules for calculating (*a*) sums and differences of integers; (*b*) products of integers.

2. Give a geometric interpretation for $|x - y|$, where "x" and "y" denote real numbers.

3. Graph the equations:

a. $y = |x|$
b. $y = |x + 2|$
c. $y = |x - 2|$
d. $y = |x^2 - 3x + 2|$

4. Describe informally the relation of the graph of $y = f(x)$ to the graph of $y = |f(x)|$.

6.16 INEQUALITIES AND ABSOLUTE VALUE

The further the student goes in many branches of mathematics, the more important the ability to handle inequalities becomes. For example, in elementary calculus one needs the theorem

$$\lim_{x \to 0} \frac{\sin x}{x} = 1. \tag{6.43}$$

In the language of symbolic logic (6.43) can be written

$$(\epsilon)(\exists\delta)(x)\left[(\epsilon > 0)(\delta > 0)(|x| < \delta) \rightarrow \left|\frac{\sin x}{x} - 1\right| < \epsilon\right].$$

While one does not attempt a very formal proof of such a complicated statement in elementary calculus, the mathematics student must ultimately be able to prove such statements, and this requires an ability to handle inequalities involving absolute value symbols. It is just at this stage that some students reach their limit in college mathematics. It is safe to say that the more acquaintance the student has with inequalities and the concept of absolute value the better equipped he is for higher mathematics. The high school student bent on pursuing a mathematics or science program in college should certainly have the opportunity to start this acquaintance with inequalities in advanced algebra, if not earlier.

The lemmas and theorems that follow hold in any ordered field.

L_{13}: $(x)[(|x| = x)\vee(|x| = -x)]$

The proof follows easily from the definition of absolute value, O_1, and the VSF "$(P \to Q)(\sim P \to R) \to Q \vee R$", but we give a proof, without assuming the VSF, which is typical of the *proofs by cases* needed for dealing with inequalities.

1. $(x \geq 0 \to \lvert x\rvert = x)(x < 0 \to \lvert x\rvert = -x)$	1. IU, def of absol val
2. $x \geq 0$	2. Assumption
3. $\lvert x\rvert = x$	3. Simp, Step 1, mod pon [2]
4. $\lvert x\rvert = x \to (\lvert x\rvert = x) \vee (\lvert x\rvert = -x)$	4. VSF $\{P \to P \vee Q\}$
5. $(\lvert x\rvert = x) \vee (\lvert x\rvert = -x)$	5. Mod pon [3,4]
6. $x \geq 0 \to (\lvert x\rvert = x) \vee (\lvert x\rvert = -x)$	6. Ded prin [Assump 2 disch'd]
7. $x < 0$	7. Assumption
8. $\lvert x\rvert = -x$	8. Simp, Step 1, mod pon [7]
9. $\lvert x\rvert = -x \to (\lvert x\rvert = x) \vee (\lvert x\rvert = -x)$	9. VSF $\{P \to Q \vee P\}$
10. $(\lvert x\rvert = x) \vee (\lvert x\rvert = -x)$	10. Mod pon [8,9]
11. $x < 0 \to (\lvert x\rvert = x) \vee (\lvert x\rvert = -x)$	11. Ded prin [Assump 7 disch'd]
12. $(x \geq 0) \vee (x < 0) \to (\lvert x\rvert = x) \vee (\lvert x\rvert = -x)$	12. Inf by cases [6,11]
13. $(x \geq 0) \vee (x < 0)$	13. IU, O_1
14. $(\lvert x\rvert = x) \vee (\lvert x\rvert = -x)$	14. Mod pon [13,12]
15. L_{13}	15. PGU

The next three lemmas are proved in much the same way; the proofs are about the same length, and are left as exercises.

L_{14}: $(x)[\lvert x\rvert \geq 0]$

L_{15}: *a.* $(x)[\lvert x\rvert \geq x]$

b. $(x)[\lvert x\rvert \geq -x]$

The next theorem and its corollaries state important equivalences:

T_{23}: $(x,a)[(a > 0)(-a < x < a) \leftrightarrow \lvert x\rvert < a]$

1. $(a > 0)(-a < x < a)$	1. Assumption
2. $(x \geq 0 \to \lvert x\rvert = x)(x < 0 \to \lvert x\rvert = -x)$	2. IU, def absol val
3. $x \geq 0$	3. Assumption
4. $\lvert x\rvert = x$	4. Simp, Step 2, mod pon [3]
5. $x < a$	5. Simp, Step 1
6. $\lvert x\rvert < a$	6. IS [4,5]
7. $x \geq 0 \to \lvert x\rvert < a$	7. Ded prin [disch's assump 3]

8. $x < 0$	8. Assump
9. $\lvert x\rvert = -x$	9. Simp, Step 2, mod pon [8]
10. $-a < x$	10. Simp, Step 1
11. $-a < x \rightarrow -x < -(-a)$	11. IU, T_{13}
12. $-x < -(-a)$	12. Mod pon [10,11]
13. $a = -(-a)$	13. IU, T_{10}
14. $-x < a$	14. IS [13,12]
15. $\lvert x\rvert < a$	15. IS [9,14]
16. $x < 0 \rightarrow \lvert x\rvert < a$	16. Ded prin (Assump 8 disch'd]
17. $(x \geq 0)\vee(x < 0) \rightarrow \lvert x\rvert < a$	17. Inf by cases [7,16]
18. $(x \geq 0)\vee(x < 0)$	18. IU, O_1
19. $\lvert x\rvert < a$	19. Mod pon [18,17]
20. $(1) \rightarrow (19)$	20. Ded prin

This proves half of the desired equivalence. The other half is left as an exercise.

Cor 1: $(x,a,b)[(b > 0)(a - b < x < a + b) \leftrightarrow |x - a| < b]$

To prove the corollary, first find a substitution instance for T_{23} by substituting "$x - a$" for "x" and "b" for "a".

Cor 2: $(x,a)[(a \geq 0)(-a \leq x \leq a) \rightarrow |x| \leq a]$

T_{24}: $(x,a)\{a > 0 \rightarrow [x^2 \leq a^2 \rightarrow |x| \leq a]\}$

Proof: (contrapositive form)

1. $a > 0$	1. Assumption
2. $\lvert x\rvert > a$	2. Assumption
3. $(\lvert x\rvert > a)(a > 0) \rightarrow \lvert x\rvert > 0$	3. IU, O_3 [using the abbrev. "$>$"]
4. $\lvert x\rvert > 0$	4. Mod pon [2,1,3]
5. $\lvert x\rvert > 0 \rightarrow [\lvert x\rvert > a \rightarrow \lvert x\rvert^2 > a\lvert x\rvert]$	5. IU, O_5 [abbrev. "$>$"]
6. $\lvert x\rvert > a \rightarrow \lvert x\rvert^2 > a\lvert x\rvert$	6. Mod pon [5,4]
7. $\lvert x\rvert^2 > a\lvert x\rvert$	7. Mod pon [6,2]
8. $a > 0 \rightarrow [\lvert x\rvert > a \rightarrow a\lvert x\rvert > a^2]$	8. IU, O_5
9. $\lvert x\rvert > a \rightarrow a\lvert x\rvert > a^2$	9. Mod pon [1,8]
10. $a\lvert x\rvert > a^2$	10. Mod pon [2,9]
11. $(\lvert x\rvert^2 > a\lvert x\rvert)(a\lvert x\rvert > a^2) \rightarrow \lvert x\rvert^2 > a^2$	11. IU, O_3
12. $\lvert x\rvert^2 > a^2$	12. Mod pon [7,10,11]
13. $\lvert x\rvert^2 = x^2$	13. IU, Exercise 12, page 233

14. $x^2 > a^2$	14. IS [13,12]
15. $\|x\| > a \to x^2 > a^2$	15. Ded prin [Assump 2 disch'd]
16. $\sim(x^2 > a^2) \to \sim(\|x\| > a)$	16. Subst. VSF [contrapositive]
17. $\sim(x^2 > a^2) \leftrightarrow x^2 \leq a^2$	17. IU, T_{17}
18. $x^2 \leq a^2 \to \sim(\|x\| > a)$	18. Inf by hyp syll
19. $\sim(\|x\| > a) \leftrightarrow \|x\| \leq a$	19. IU, T_{17}
20. $x^2 \leq a^2 \to \|x\| \leq a$	20. Inf by hyp syll
21. $a > 0 \to (20)$	21. Ded prin [1,20]
22. T_{24}	22. PGU

Cor 1: $(x,a)\{a > 0 \to [x^2 > a^2 \to |x| > a]\}$
Cor 2: $(x,a)\{a > 0 \to [x^2 \geq a^2 \to |x| \geq a]\}$
Cor 3: $(x,a)\{a > 0 \to [x^2 = a^2 \to |x| = a]\}$

In the field of real numbers, the symbol "$\sqrt{\ }$" is so defined that if "$a \geq 0$" then "$\sqrt{a} \geq 0$" is true. For example, $\sqrt{4} = 2$ and $\sqrt{4} \neq -2$. Hence, in the real number interpretation, T_{24} can be written

$$(x,a)\{a > 0 \to [x^2 \leq a \to |x| \leq \sqrt{a}]\}.$$

Indeed, an even simpler version is

$$(x,a)[x^2 \leq a \to |x| \leq \sqrt{a}].$$

With T_{23}, T_{24}, and the above result, several types of inequality problems associated with quadratic equations can be dealt with as easily as one solves simple equations. For example,

PROBLEM: Find the necessary and sufficient conditions on the real coefficients a,b,c with $a \neq 0$, so that the roots of $ax^2 + bx + c = 0$ are imaginary.

We assume it known that the roots are

$$\frac{-b + \sqrt{b^2 - 4ac}}{2a}, \quad \frac{-b - \sqrt{b^2 - 4ac}}{2a}.$$

Assume both roots imaginary. Then

$$b^2 - 4ac < 0.$$

Or, using T_{16} or O_4,

$$b^2 < 4ac.$$

Now, since $0 \leq b^2$ (easily from T_{14}), it follows that $4ac > 0$ and T_{23}, Cor. 1, applies to yield

$$|b| < 2\sqrt{ac}.$$

On the other hand, if we like, using T_{22}

$$-2\sqrt{ac} < b < 2\sqrt{ac}$$

is the necessary condition.

To prove the condition sufficient, one would have to reverse all steps. This can be done using T_{22} and T_{23}, Cor. 3.

EXERCISES

1. Prove L_{14}, page 230.
2. Prove $L_{15}(a)$, page 230.
3. Prove $L_{15}(b)$, page 230.
4. Prove the other half of T_{23}, page 230.
5. Prove T_{23} Cor. 1, page 231.
6. Prove T_{23} Cor. 2, page 231.
7. Prove T_{24} Cor. 1, page 232.
8. Prove T_{24} Cor. 2, page 232.
9. Prove T_{24} Cor. 3, page 232.
10. Prove: $(x)[|x| = 0 \rightarrow x = 0]$.
11. Prove: $(x)[|x| = |-x|]$.
12. Prove: $(x)[x^2 = |x|^2]$.
13. Prove: $|0| = 0$.
14. Prove: $(x)[x^2 \geq 0]$ (very short proof).
15. Prove: $(x,y)[x^2 + y^2 \geq 2xy]$ (HINT: Use T_{14}).
16. Abbreviate the proof of T_{24}, and then write it out without using the symbolic language.
17. Prove: $|ab| = |a||b|$.

Early in the calculus, one needs the relations

$$|ab| = |a||b|$$
$$|a + b| \leq |a| + |b|.$$

The equality is easy to establish and is left as an exercise. The inequality follows from L_{14}, L_{15}, T_{18}, T_{23}:

T_{25}: $(x,y)[|x + y| \leq |x| + |y|$

1. $\|x\| \geq x$	1. IU, $L_{15}(a)$
2. $\|y\| \geq y$	2. IU, $L_{15}(a)$
3. $(\|x\| \geq x)(\|y\| \geq y) \rightarrow$ $\|x\| + \|y\| \geq x + y$	3. IU, T_{18} Cor.
4. $\|x\| + \|y\| \geq x + y$	4. Mod pon [1,2,3]
5. $\|x\| + \|y\| \geq -(x + y)$	5. Similarly from $L_{15}(b)$

6. $(5) \to -(\lvert x\rvert + \lvert y\rvert) \leq x + y$	6. IU, T_{13}, T_{17}
7. $-(\lvert x\rvert + \lvert y\rvert) \leq x + y$	7. Mod pon [6,5]
8. $-(\lvert x\rvert + \lvert y\rvert) \leq x + y \leq \lvert x\rvert + \lvert y\rvert$	8. Conj inf [7,4]
9. $\lvert x\rvert \geq 0$	9. IU, L_{14}
10. $\lvert y\rvert \geq 0$	10. IU, L_{14}
11. $(9)(10) \to \lvert x\rvert + \lvert y\rvert \geq 0$	11. IU, T_{18} Cor.
12. $\lvert x\rvert + \lvert y\rvert \geq 0$	12. Mod pon [9,10,11]
13. $(12)(8) \to \lvert x + y\rvert \leq \lvert x\rvert + \lvert y\rvert$	13. IU, T_{23}
14. $\lvert x + y\rvert \leq \lvert x\rvert + \lvert y\rvert$	14. Mod pon [8,12,13]
15. T_{25}	15. PGU

6.17 APPLICATIONS TO NOTIONS OF LIMIT

Sometimes a student first encounters the notion of a limit briefly in plane geometry. There may be further brief encounters in trigonometry and analytic geometry, but it is in calculus that the notion is first systematically exploited. From this time on, as long as he studies mathematics, his notion of limit will broaden and deepen. In the early stages, calculus is mainly concerned with limiting processes involving real numbers, and makes heavy use of the ordered field properties of the foregoing sections, although further properties of the real numbers are also needed.

A basic notion is that of *limit of a sequence.* For this discussion, a sequence is a real valued function defined over all the positive integers. We symbolize a real sequence by "$\{A_n\}$", where "A_n" denotes the value of the function for the positive integer n.

A particular sequence can be described in many ways, but most commonly by showing how to compute A_n for each n; for example (in this example, and in what follows, "n" is reserved to denote positive integers),

$$\{A_n\} \text{ is the sequence such that } (n)[A_n = n^2].$$

This sequence has for its first five terms

$$1, 4, 9, 16, 25.$$

By the limit A of a sequence $\{A_n\}$, we mean, loosely, a real number A such that all but a finite number of terms of the sequence are arbitrarily close to A. More precisely

$$\lim A_n = A \underset{\text{def}}{\leftrightarrow} (\epsilon)(\exists N)\{\epsilon > 0 \to (n)[n > N \to |A_n - A| < \epsilon]\}$$

where we limit the values for the variables "n" and "N" to positive integers.

In this definition, ϵ is a measure of the closeness of A_n to A, and N is a measure of how far out to go in the sequence so that terms beyond A_N will be close enough to A. Let us prove that if

$$A_n = \frac{2^n - 1}{2^n},$$

then $\lim A_n = 1$. Here, we must make $|A_n - A|$, or in this case

$$\left|\frac{2^n - 1}{2^n} - 1\right|,$$

small. Now

$$\left|\frac{2^n - 1}{2^n} - 1\right| = \left|\frac{2^n - 1 - 2^n}{2^n}\right| = \left|\frac{-1}{2^n}\right| = \frac{1}{2^n}$$

and

$$\frac{1}{2^n} < \epsilon \leftrightarrow \frac{1}{\epsilon} < 2^n$$
$$\frac{1}{\epsilon} < 2^n \leftrightarrow \log\frac{1}{\epsilon} < \log 2^n$$
$$\log\frac{1}{\epsilon} < \log 2^n \leftrightarrow \log\frac{1}{\epsilon} < n \log 2$$
$$\log\frac{1}{\epsilon} < n \log 2 \leftrightarrow \frac{\log 1/\epsilon}{\log 2} < n.$$

So for arbitrary $\epsilon > 0$, we should choose

$$N > \frac{\log 1/\epsilon}{\log 2}.$$

It is a property of the real numbers that any real number is exceeded by some integer. We now have the ingredients to prove that the limit of the sequence is 1.

1. $\epsilon > 0$	1. Assump
2. $(\exists N)\left(N > \frac{\log 1/\epsilon}{\log 2}\right)$	2. Property of real numbers and integers
3. $N > \frac{\log 1/\epsilon}{\log 2}$	3. IE
4. $n > N$	4. Assump
5. $(3)(4) \rightarrow n > \frac{\log 1/\epsilon}{\log 2}$	5. IU, O_3
6. $n > \frac{\log 1/\epsilon}{\log 2}$	6. Mod pon [3,4,5]
7. $(6) \rightarrow \left\|\frac{2^n - 1}{2^n} - 1\right\| < \epsilon$	7. See preliminary discussion

8. $\left\lvert \frac{2^n - 1}{2^n} - 1 \right\rvert < \epsilon$	8. Mod pon
9. $n > N \rightarrow \left\lvert \frac{2^n - 1}{2^n} - 1 \right\rvert < \epsilon$	9. Ded prin [disch's assump 4]
10. $(n)(9)$	10. PGU
11. $\epsilon > 0 \rightarrow (n)\left[n > N \rightarrow \left\lvert \frac{2^n - 1}{2^n} - 1 \right\rvert < \epsilon\right]$	11. Ded prin [disch's assump 1]
12. $(\exists N)(11)$	12. PGE
13. $(\epsilon)(\exists N)\left\{\epsilon > 0 \rightarrow (n)\left[n > N \rightarrow \left\lvert \frac{2^n - 1}{2^n} - 1 \right\rvert < \epsilon\right]\right\}$	13. PGU

This proof is comparatively informal. Much of the abbreviation occurs at Step 7. To establish Step 7 formally would involve formalizing the preliminary discussion by means of theorems and axioms of foregoing sections. It is worth noting that it is just Step 7 that is peculiar to the given sequence, and really is the essence of the proof. The rest of the steps are common to all such proofs of the limit of a sequence. It is also worth commenting upon that it is just the relation of Step 7 to the rest of the proof that seems to cause the most trouble for beginning students.

As a last example, we present a standard proof that makes use of T_{25}.

If $\lim a_n = A$ and $\lim b_n = B$, then $\lim (a_n + b_n) = A + B$.

What we need to make arbitrarily small here is

$$|(a_n + b_n) - (A + B)|,$$

which may be written

$$|(a_n - A) + (b_n - B)|,$$

and, by T_{25},

$$|(a_n - A) + (b_n - B)| \leq |a_n - A| + |b_n - B|.$$

Since both terms on the right of this inequality can be made arbitrarily small by definition and the hypotheses, we can see our way clear to a proof:

1. $(\epsilon)(\exists N)\{\epsilon > 0 \rightarrow (n)[n > N \rightarrow \lvert a_n - A \rvert < \epsilon]\}$	1. Given
2. $(\epsilon)(\exists N)\{\epsilon > 0 \rightarrow (n)[n > N \rightarrow \lvert b_n - B \rvert < \epsilon]\}$	2. Given
3. $(\exists N)\left\{\frac{\alpha}{2} > 0 \rightarrow (n)\left[n > N \rightarrow \lvert a_n - A \rvert < \frac{\alpha}{2}\right]\right\}$	3. IU, Step 1

4. $\frac{\alpha}{2} > 0 \to (n)\left[n > N_1 \to |a_n - A| < \frac{\alpha}{2}\right]$	4. IE, Step 3
5. $\alpha > 0$	5. Assump
6. $\frac{1}{2} > 0 \to \left[\alpha > 0 \to \frac{\alpha}{2} > \frac{1}{2} \cdot 0\right]$	6. IU, O_5
7. $\frac{1}{2} > 0$	7. Property real numbers
8. $\alpha > 0 \to \frac{\alpha}{2} > \frac{1}{2} \cdot 0$	8. Mod pon
9. $\frac{\alpha}{2} > \frac{1}{2} \cdot 0$, i.e., $\frac{\alpha}{2} > 0$	9. Mod pon, and T_7
10. $(n)\left[n > N_1 \to |a_n - A| < \frac{\alpha}{2}\right]$	10. Mod pon [9,4]
11. $n > N_1 \to |a_n - A| < \frac{\alpha}{2}$	11. IU, Step 10
12. $n > N_2 \to |b_n - B| < \frac{\alpha}{2}$	12. Similarly, Steps 1, 3 to 11
13. $(\exists N)[(N > N_1)(N > N_2)]$	13. Property of the integers
14. $(N_3 > N_1)(N_3 > N_2)$	14. IE, Step 13
15. $n > N_3$	15. Assump
16. $(n > N_3)(N_3 > N_1) \to n > N_1$	16. IU, O_3
17. $n > N_1$	17. Mod pon, simp [14,15,16]
18. $n > N_2$	18. Similarly
19. $|a_n - A| < \frac{\alpha}{2}$	19. Mod pon [17,11]
20. $|b_n - B| < \frac{\alpha}{2}$	20. Mod pon [18,12]
21. $|(a_n + b_n) - (A + B)| = |(a_n - A) + (b_n - B)|$	21. Algebra
22. $|(a_n - A) + (b_n - B)| \leq |a_n - A| + |b_n - B|$	22. IU, T_{25}
23. $|(a_n + b_n) - (A + B)| \leq |a_n - A| + |b_n - B|$	23. IS [21,22]
24. $|(a_n + b_n) - (A + B)| \leq \frac{\alpha}{2} + \frac{\alpha}{2} = \alpha$	24. IS [19,20 in 23]
25. $n > N_3 \to (24)$	25. Ded prin [Assump 15 disch'd]
26. $(n)[n > N_3 \to (24)]$	26. PGU
27. $\alpha > 0 \to (26)$	27. Ded prin [Assump 5 disch'd]

28. $(\exists N)\{\alpha > 0 \rightarrow (n)[n > N \rightarrow (24)]\}$ — 28. PGE

29. $(\epsilon)(\exists N)\{\epsilon > 0 \rightarrow (n)[n > N \rightarrow |(a_n + b_n) - (A + B)| < \epsilon]\}$ — 29. PGU

Observe that because of restrictions surrounding IE it is necessary to choose N_1 and N_2 independently in Step 4 leading to 11, and the corresponding step (tacit) that leads similarly to Step 12.

EXERCISES

1. Prove: $\lim b_n = 0$, where $b_n = \frac{1}{n^2}$
2. Prove: $\lim \left[\frac{2^n - 1}{2^n} + \frac{1}{n^2}\right] = 1$

6.18 RESTRICTED QUANTIFICATION

The symbols used for individual variables in mathematics are always used in a restricted way. The "x" and "y" of algebra are generally restricted to values in some class of numbers, usually the real or the complex numbers. The statement

$$(x,y)[(x + y)^2 = x^2 + 2xy + y^2]$$

is taken to assert the truth of all substitution instances, with the understanding that symbols substituted for the variables are numerals, not names for monkeys or other such objects. Sometimes one asserts that the universe of individuals consists of real numbers, or of complex numbers, and so on. It is very common, however, to make statements whose substitution instances involve several different classes. In algebra, for instance, one might write

$$(a,b,c)(\exists x)[ax^2 + bx + c = 0],$$

where it is desired that the coefficients a,b,c are real numbers, but x may be complex.

In plane geometry, most theorems involve statements about points and lines. It is very convenient to make the notational agreement that certain letters will be used to denote points, and certain other letters will be used to denote lines. This constitutes a valuable agreement, and is handled in the logic by a formal agreement.

Suppose one wished a formal expression for the general statement

(6.44) If a point is on a line, then the line is on the point.

We take for translations:

$$P(x): x \text{ is a point.}$$
$$L(x): x \text{ is a line.}$$
$$W(x,y): x \text{ is on } y.$$

Then the translation for (6.44) is written

$$(x,y)\{P(x)L(y) \rightarrow [W(x,y) \leftrightarrow W(y,x)]\},$$

which may also be written:

(6.45) $$(x,y)\{P(x) \rightarrow [L(y) \rightarrow (W(x,y) \leftrightarrow W(y,x))]\}.$$

In any development of geometry, every axiom or theorem will involve the predicates "P" and "L". It will be a great convenience to make the following abbreviation agreement (where "F" is some unspecified predicate):

Agreement for Restricted Quantification:[1]

(6.46)
a. $(a)F(a)$ for $(x)[P(x) \rightarrow F(x)]$
b. $(\alpha)F(\alpha)$ for $(x)[L(x) \rightarrow F(x)]$,

where lower-case Latin letters, early in the alphabet, are to be reserved for abbreviation (*a*), and lower-case Greek letters are to be reserved for abbreviation (*b*).

With agreement (6.46*a*) we may abbreviate (6.45) as

$$(a,y)[L(y) \rightarrow (W(a,y) \leftrightarrow W(y,a))],$$

and, using agreement (6.46*b*), as

(6.47) $$(a,\alpha)[W(a,\alpha) \leftrightarrow W(\alpha,a)].$$

The formal advantages of (6.47) over (6.45) are clear.

For existential quantification we agree[2]

(6.46)
c. $(\exists a)F(a)$ for $(\exists x)[P(x)F(x)]$
d. $(\exists \alpha)F(\alpha)$ for $(\exists x)[L(x)F(x)]$.

With these agreements, we may translate

on every line is at least one point,

as

(6.48) $$(\alpha)(\exists a)[W(a,\alpha)].$$

[1] Compare the formula with our earlier translations for the Aristotelian categorical propositions, 5.18.

[2] Again, compare with 5.18.

Without the agreements (6.46) we would have to write (6.48) as

$$(x)(\exists y)[L(x) \rightarrow P(y)W(y,x)],$$

or perhaps,

$$(x)\{L(x) \rightarrow (\exists y)[P(y)W(y,x)]\}.$$

So far, we have made agreements for abbreviating quantified statements. We must examine how IU and IE behave with these agreements. Clearly, by IU, from

$$(x)[P(x) \rightarrow F(x)]$$

we can get

$$P(y) \rightarrow F(y).$$

It is natural to expect to apply IU to

$$(a)F(a)$$

and obtain

$$F(b). \tag{6.49}$$

However, one would also expect to apply IE to

$$(\exists a)F(a)$$

to obtain

$$F(b), \tag{6.50}$$

with, of course, the usual restrictions on b in future applications of PGU. The difficulty here is that (6.49) is an abbreviation for

$$P(y) \rightarrow F(y); \tag{6.51}$$

and (6.50) is an abbreviation for

$$P(y)F(y). \tag{6.52}$$

Now, (6.51) and (6.52) are quite different. If one is to use IU and IE in this way, it seems that one will have to remember where such a formula as "$F(a)$" comes from, in order to know what its unabbreviated form is. Observe that, while it is a simple thing to prove

$$(x)F(x) \rightarrow (\exists x)F(x),$$

it is not possible to prove

$$(a)F(a) \rightarrow (\exists a)F(a). \tag{6.53}$$

Let us try, using unabbreviated forms. We want to demonstrate

$$(x)[P(x) \rightarrow F(x)] \rightarrow (\exists x)[P(x)F(x)]. \tag{6.54}$$

1. $(x)[P(x) \rightarrow F(x)]$ — 1. Assumption
2. $P(y) \rightarrow F(y)$ — 2. IU, Step 1

At this point we are stuck, because there is no way of deducing "$P(y)$", which is necessary at this point. Suppose, however, that there is an axiom of the geometry stating the existence of points,

Axiom: $(\exists x)P(x)$.

Then it is possible to prove (6.54) as follows:

1. $(x)[P(x) \to F(x)]$	1. Assumption
2. $(\exists x)P(x)$	2. Axiom
3. $P(y)$	3. IE, Step 2
4. $P(y) \to F(y)$	4. IU, Step 1
5. $F(y)$	5. Mod pon [3,4]
6. $P(y)F(y)$	6. Conj inf [3,5]
7. $(\exists x)[P(x)F(x)]$	7. PGE
8. $(x)[P(x) \to F(x)] \to (\exists x)[P(x)F(x)]$	8. Ded prin

Thus it is not possible to prove (6.53) without the statement that points exist in the geometry. Of course, such a statement will be either an axiom or a theorem of the geometry, so that statements like (6.53) can be proved, and we may operate with restricted quantification in much the same way as we have before these abbreviation agreements.

It remains to discuss the meaning to be attached to "$F(a)$" when it is not derived by IU or IE. The situation will arise when "$F(a)$" is stated as an assumption. We shall agree that whenever "$F(a)$" is assumed (later to be discharged from the demonstration) it is an abbreviation for "$P(x)F(x)$", and that "$G(\alpha)$" so assumed is an abbreviation for "$L(x)G(x)$".

If one arrives at "$F(a)$" at a step in a demonstration, before applying PGE, one must check to see that "$P(x)F(x)$" could have been demonstrated. This generally involves looking back to see how "a" was introduced. Before applying PGU, one checks all the usual restrictions, checks to see that "$P(x) \to F(x)$" could have been demonstrated, before proceeding from "$F(a)$" to "$(a)F(a)$". In general, if all the other restrictions for PGU are met, it will be easy to verify this additional point by looking back to see how "a" was first introduced into the demonstration. The common situation is that the first step in the demonstration was something like

1. $G(a)$, that is, $P(x)G(x)$ 1. Assumption

Then at some later step one has demonstrated

$$F(a).$$

Then by the deduction principle

$$G(a) \rightarrow F(a),$$

which without abbreviation would be written

$$P(x)G(x) \rightarrow F(x).$$

Using the valid statement formula "$[PG \rightarrow F] \leftrightarrow [P \rightarrow (G \rightarrow F)]$", we may write

$$P(x) \rightarrow [G(x) \rightarrow F(x)];$$

then, by PGU,

$$(x)\{P(x) \rightarrow [G(x) \rightarrow F(x)]\};$$

and this may be abbreviated

$$(a)[G(a) \rightarrow F(a)].$$

The first lemma of the Appendix is an example of this pattern.

While we have presented these notions of restricted quantification in a context of geometry, they may be used wherever it is desired to deal with several classes of objects in an abstract system.

Appendix

SYMBOLIC TREATMENT OF THE MINIATURE GEOMETRY

With the machinery that has been developed, it is possible to give a fairly rigorous treatment of the miniature geometry presented in Sec. 4.2. What follows is a purely formal exercise and constitutes a test of the adequacy for geometry of the formal machinery so far developed. The formal problems in developing the geometry are more complex than those we faced in developing the abstract field system, as comparison of this Appendix with Chap. VI will show. On the other hand, comparison of the Appendix with Sec. 4.2 will bring home just how informal that section really is. We believe that anyone making these comparisons will agree that algebra is a far simpler, and more effective, system for teaching axiomatic structure than is geometry—even miniature geometry.

Some of those readers who have come this far with us may wish to try their own hands at a formal development of Sec. 4.2, before reading further. We suggest that it will be necessary to make full use of the methods of Sec. 6.18 in such a project.

Certain informalities are readily apparent in the demonstrations that follow. We list some of them and hope that such other informalities as occur will be readily understood:

1. Associative and commutative properties for conjunction and disjunction of statement functions will be exploited without specific mention.
2. Substitutions from valid statement formulas will occur without any appearance of the valid statement formula as a step in the demonstration. When this happens, the analysis column will have in it the expression "Subst. VSF".
3. Within a demonstration, where a long sequence of steps would essentially duplicate a sequence of steps already existing, just the conclusion is written, and the analysis column will read "Similarly" followed by an exact reference to the similar steps.
4. Where indirect subdemonstrations occur, the indirect assumption will be indicated by "Assump" in the analysis column. When a contradiction is reached, it will be followed directly by "CONTRA", and the next step will be a statement of the negation of the assumption. It is understood that such a procedure discharges the assumption.

Many of the demonstrations that follow are quite long, but it should be borne in mind that much of the length results from the care taken to prove, wherever necessary, that the points and lines under discussion are distinct. No special effort has been made to find the shortest demonstrations, and no doubt all the longer demonstrations can be substantially shortened. Rather we tried to develop the geometry along the lines of Sec. 4.2 when practical. There are, however, differences in the organization of the axioms and in numbering theorems and lemmas.

First, we restate the axioms in a form that will be easy to translate into symbolic language. We restate the axioms in such a way as to avoid the word "exactly", since our formalism has no direct means of translating this notion.

REPHRASING OF THE AXIOMS

A_0: For every x and y, if x is on y then y is on x.

NOTE: In the geometry it will always happen that one of "x", "y" will be interpreted as a point and the other as a line.

A_1: There is at least one line.
$A_{2.1}$: Every line has at least three distinct points on it.
$A_{2.2}$: If there are four points on a line, then they are not all distinct.

Definition: (a is a pole of α) For every line β and point b, it is not the case that a is joined to b on α by β.

Definition: (α is a polar of a) For every point b and line β, it is not the case that α is joined to β on a by b.

$A_{3.1}$: Every line has a pole.
$A_{3.2}$: If a and b are poles of the same line, then a = b.
$A_{4.1}$: Every point has a polar.
$A_{4.2}$: If α and β are polars of the same point, then $\alpha = \beta$.
$A_{5.1}$: If a is not on α, but is joined to b on α by a line β, then a is also joined to another point c on α by a line γ.
$A_{5.2}$: If a is not on α, then there is a point on α that is not joined to a by any line.

SPECIAL SYMBOLS AND ABBREVIATIONS

We are now ready to start the formal development. First, we name the special symbols.

Individual Variables:

$x, y, z, w, t, \ldots$ (from the end of the alphabet)

Functional Constants	*Interpretation*
L	$L(x)$: x is a line.
P	$P(x)$: x is a point.
W	$W(x,y)$: x is on y.

We shall make full use of restricted quantification; see Sec. 6.18. Some of the abbreviations follow.

Formula	*Abbreviation*	*Remarks*
$(x)[L(x) \to F(x)]$	$(\alpha)F(\alpha)$	Small Greek letters are reserved
$(\exists x)[L(x)F(x)]$	$(\exists\beta)F(\beta)$	for these two abbreviations.
$(x)[P(x) \to G(x)]$	$(a)G(a)$	Small Latin letters, other than
$(\exists x)[P(x)G(x)]$	$(\exists b)G(b)$	those already reserved for individual variables, will be used for these abbreviations.
$(x,y)[P(x) \to (L(x) \to W(x,y))]$	$(a,\beta)W(a,\beta)$	
$(x,y)[P(x) \to (\exists y)(L(y)W(x,y))]$	$(a)(\exists\beta)W(a,\beta)$	

Special Abbreviations:

$D(x,y,z)$ for $(x \neq y)(y \neq z)(z \neq x)$
$D(x,y,z,w)$ for $D(x,y,z)(x \neq w)(y \neq w)(z \neq w)$
etc.
$J(x,y;z)$ for $W(x,z)W(y,z)$
$J(x,y,z;w)$ for $W(x,w)W(y,w)W(z,w)$ or $J(x,y;w)W(z,w)$

NOTE: "$J(x,y;z)$" is read "x and y are joined by z". It is clear that "$J(x,y;z) \leftrightarrow J(y,x;z)$" is true, and we shall use this fact repeatedly without further mention.

$J(x;y,z)$ for $W(x,z)W(y,z)$
$J(x;y,z,w)$ for $J(x;y,z)W(x,w)$

THE AXIOMS

A_0: $(x,y)[W(x,y) \leftrightarrow W(y,x)]$ [A_0 will be used tacitly in general.]
A_1: $(\exists x)L(x)$
$A_{2.1}$: $(\alpha)(\exists b,c,d)[J(b,c,d;\alpha)D(b,c,d)]$
$A_{2.2}$: $(\alpha,b,c,d,e)[J(b,c,d,e;\alpha) \rightarrow \sim D(b,c,d,e)]$

Abbreviations:

Poles: $H(a,\alpha)$ for $(\beta,b)\sim[W(b,\alpha)J(a,b;\beta)]$
Polars: $G(\alpha,a)$ for $(b,\beta)\sim[W(a,\beta)J(b;\alpha,\beta)]$

$A_{3.1}$: $(\alpha)(\exists a)H(a,\alpha)$
$A_{3.2}$: $(\alpha,a,b)[H(a,\alpha)H(b,\alpha) \rightarrow a = b]$
$A_{4.1}$: $(a)(\exists\alpha)G(\alpha,a)$
$A_{4.2}$: $(a,\alpha,\beta)[G(\alpha,a)G(\beta,a) \rightarrow \alpha = \beta]$
$A_{5.1}$: $(a,\alpha,b)\{\sim W(a,\alpha)W(b,\alpha)(\exists\beta)J(a,b;\beta) \rightarrow$
$(\exists c,\gamma)[W(c,\alpha)J(a,c;\gamma)(b \neq c)]\}$
$A_{5.2}$: $(a,\alpha)\{\sim W(a,\alpha) \rightarrow (\exists b)[W(b,\alpha)(\beta)\sim J(a,b;\beta)]\}$

LEMMAS AND THEOREMS

L_1: $(a,\alpha)[H(a,\alpha) \rightarrow \sim W(a,\alpha)]$

Proof:

1. $H(a,\alpha)$	1. Assumption
2. $\sim[W(a,\alpha)J(a,a;\alpha)]$, i.e., $\sim[W(a,\alpha)W(a,\alpha)W(a,\alpha)]$	2. IU, 1 [see def. $H(a,\alpha)$]
3. $W(a,\alpha) \rightarrow W(a,\alpha)W(a,\alpha)W(a,\alpha)$	3. VSF [$A \rightarrow AAA$]
4. $\sim W(a,\alpha)$	4. Contrapos inf 3,2
5. $H(a,\alpha) \rightarrow \sim W(a,\alpha)$	5. Ded prin [disch assump 1]
6. $(a,\alpha)[H(a,\alpha) \rightarrow \sim W(a,\alpha)]$	6. PGU [applied twice]

Note that "a" and "α" in assumption 1 are free; hence by our agreement, the unabbreviated assumption is "$P(x)L(y)H(x,y)$". Therefore (5) unabbreviated is "$P(x)L(y)H(x,y) \rightarrow \sim W(x,y)$", which may also be written in the form "$P(x) \rightarrow \{L(x) \rightarrow [H(x,y) \rightarrow \sim W(x,y)]\}$". Thus after generalizing on "x" and "y" by PGU, the result is correctly abbreviated as in (6).

L_2: $(a,\alpha)[G(\alpha,a) \rightarrow \sim W(a,\alpha)]$

The proof is similar to that of L_1, and is left as an exercise.

L_3: $(a,\alpha)\{\sim H(a,\alpha) \rightarrow (\exists\beta,b)[W(b,\alpha)J(a,b;\beta)]\}$

Proof:

1. $\sim H(a,\alpha)$, i.e., $\sim(\beta,b)[W(b,\alpha)J(a,b;\beta)]$	1. Assumption
2. $(\exists\beta,b)\sim\sim[W(b,\alpha)J(a,b;\beta]$	2. Quant conv [see Sec. 5.4]
3. $\sim\sim[W(d,\alpha)J(a,d;\gamma)]$	3. IE (twice) 2
4. $W(d,\alpha)J(a,d;\gamma)$	4. Subst. VSF
5. $(\exists\beta,b)[W(b,\alpha)J(a,b;\beta)]$	5. PGE (twice) 4
6. $\sim H(a,\alpha) \rightarrow (5)$	6. Ded prin 1,5
7. L_3	7. PGU (twice) 6

L_4: $(a,\alpha)\{(\exists b,c)[J(b,c;\alpha)(b \neq c)(\beta)\sim J(a,b;\beta)(\gamma)\sim J(a,c;\gamma)] \rightarrow H(a,\alpha)\}$

Proof: [Indirect. Prove "$P \rightarrow Q$" by showing that "$P\sim Q$" leads to contradiction.]

1. $(\exists b,c)[J(b,c;\alpha)(b \neq c)(\beta) \sim J(a,b;\beta)$ $(\gamma)\sim J(a,c;\gamma)]\sim H(a,\alpha)$	1. Assumption
2. $\sim H(a,\alpha)$	2. Simp 1
3. $\sim H(a,\alpha) \rightarrow (\exists\beta,b)[W(b,\alpha)J(a,b;\beta)]$	3. IU, L_3
4. $(\exists\beta,b)[W(b,\alpha)J(a,b;\beta)]$	4. Mod pon 2,3
5. $W(d,\alpha)J(a,d;\delta)$	5. IE 4
6. $J(m,n;\alpha)(m \neq n)(\beta)\sim J(a,m;\beta)$ $(\gamma)\sim J(a,n;\gamma)\sim H(a,\alpha)$	6. IE 1
7. $J(a,d;\delta)$	7. Simp 1
8. $(\beta)\sim J(a,m;\beta)$	8. Simp 6
9. $\sim J(a,m;\delta)$	9. IU 8
10. $m \neq d$	10. ISC 7,9
11. $(\gamma)\sim J(a,n;\gamma)$	11. Simp 6
12. $\sim J(a,n;\delta)$	12. IU 11
13. $n \neq d$	13. ISC 7,12
14. $J(m,n;\alpha)$	14. Simp 6
15. $W(m,\alpha)$	15. Simp 14
16. $\sim J(a,m;\alpha)$	16. IU 8
17. $\sim[W(a,\alpha)W(m,\alpha)] \leftrightarrow$ $[W(m,\alpha) \rightarrow \sim W(a,\alpha)]$	17. VSF
18. $W(m,\alpha) \rightarrow \sim W(a,\alpha)$	18. Mod pon 16,17

19. $\sim W(a,\alpha)$ — 19. Mod pon 15,18
20. $\sim W(a,\alpha)W(d,\alpha)(\exists\beta)J(a,d;\beta) \rightarrow (\exists c,\gamma)[W(c,\alpha)J(a,c;\gamma)(d \neq c)]$ — 20. IU, $A_{5.1}$
21. $W(d,\alpha)$ — 21. Simp 5
22. $(\exists\beta)J(a,d;\beta)$ — 22. PGE 7
23. $\sim W(a,\alpha)W(d,\alpha)(\exists\beta)J(a,d;\beta)$ — 23. Conj inf 19,21,22
24. $(\exists c,\gamma)[W(c,\alpha)J(a,c;\gamma)(d \neq c)]$ — 24. Mod pon 23,20
25. $W(f,\alpha)J(a,f;\phi)(d \neq f)$ — 25. IE 24
26. $W(f,\alpha)$ — 26. Simp 25
27. $J(m,n,d,f;\alpha)$ — 27. Conj inf 14,21,26
28. $J(m,n,d,f;\alpha) \rightarrow \sim D(m,n,d,f)$ — 28. IU, $A_{2.2}$
29. $\sim D(m,n,d,f)$ — 29. Mod pon 27,28
30. $\sim J(a,m;\phi)$ — 30. IU 8
31. $J(a,f;\phi)$ — 31. Simp 25
32. $f \neq m$ — 32. ISC 30,31
33. $\sim J(a,n;\phi)$ — 33. IU 11
34. $f \neq n$ — 34. ISC 31, 33
35. $m \neq n$ — 35. Simp 6
36. $d \neq f$ — 36. Simp 25
37. $D(m,n,d,f)$ — 37. Conj inf 35,34,32,36
38. $D(m,n,d,f) \sim D(m,n,d,f)$ CONTRA — 38. Conj inf 37,29
39. $\sim(1)$ — 39. The contradiction (38) discharges the assumption (1) and establishes (39).
40. L_4 — 40. PGU (twice) 39

The demonstration of the theorem that follows is mainly an exercise in the associativity and commutativity properties of conjunction. In the future, these properties will be employed tacitly, but for this proof we put in all the steps.

T_1: $(a,\alpha)[H(a,\alpha) \leftrightarrow G(\alpha,a)]$

Proof:

1. $H(a,\alpha)$, i.e., $(b,\beta)\sim[W(b,\alpha)J(a,b;\beta)]$ — 1. Assumption
2. $\sim[W(d,\alpha)J(a,d;\delta)]$ — 2. IU 1
3. $W(d,\alpha)[W(a,\delta)W(d,\delta)] \leftrightarrow [W(d,\alpha)W(a,\delta)]W(d,\delta)$ — 3. VSF [associativity]
4. $\sim\{[W(d,\alpha)W(a,\delta)]W(d,\delta)\}$ — 4. Subst. 3 in 2
5. $W(d,\alpha)W(a,\delta) \leftrightarrow W(a,\delta)W(d,\alpha)$ — 5. VSF [commutativity]
6. $\sim\{[W(a,\delta)W(d,\alpha)]W(d,\delta)\}$ — 6. Subst. 5 in 4
7. $[W(a,\delta)W(d,\alpha)]W(d,\delta) \leftrightarrow W(a,\delta)[W(d,\alpha)W(d,\delta)]$ — 7. VSF [associativity]

8. $\sim\{W(a,\delta)[W(d,\alpha)W(d,\delta)]\}$, i.e., $\sim[W(a,\delta)J(d;\alpha,\delta)]$	8. Subst. 7 in 6
9. $(b,\beta)\sim[W(a,\beta)J(b;\alpha,\beta)]$, i.e., $G(\alpha,a)$	9. PGU 8
10. $H(a,\alpha) \rightarrow G(\alpha,a)$	10. Ded prin [disch assump 1]
11. $G(\alpha,a) \rightarrow H(a,\alpha)$	11. Proof sim. Steps 1 to 10
12. (10)(11), i.e., by def $H(a,\alpha) \leftrightarrow G(\alpha,a)$	12. Conj inf 10,11
13. T_1	13. PGU (twice)

T_2: $(a,b,\alpha,\beta)[J(a,b;\alpha)J(a,b;\beta)(a \neq b) \rightarrow \alpha = \beta]$

Proof:

1. $J(a,b;\alpha)J(a,b;\beta)(a \neq b)$	1. Assumption
2. $J(a,b;\alpha)$	2. Simp 1
3. $(\exists a)H(a,\alpha)$	3. IU, $A_{3.1}$
4. $H(c,\alpha)$	4. IE 3
5. $\sim[W(a,\alpha)J(c,a;\phi)]$	5. IU 4
6. $W(a,\alpha) \rightarrow \sim J(c,a;\phi)$	6. Subst. VSF for 5
7. $W(a,\alpha)$	7. Simp 1 [from $J(a,b;\alpha)$]
8. $\sim J(c,a;\phi)$	8. Mod pon 7,6
9. $(\beta)\sim J(c,a;\beta)$	9. PGU 8
10. $(\gamma)\sim J(c,b;\gamma)$	10. Similarly, Steps 2 to 9
11. $J(a,b;\beta)(a \neq b)$	11. Simp 1
12. $J(a,b;\beta)(a \neq b)(\beta)\sim J(c,a;\beta)(\gamma)\sim J(c,b;\gamma)$	12. Conj inf 11,9,10
13. $(\exists a,b)(12)$	13. PGE 12
14. $(13) \rightarrow H(c,\beta)$	14. IU, L_4
15. $H(c,\beta)$	15. Mod pon 13,14
16. $H(c,\beta) \leftrightarrow G(\beta,c)$	16. IU, T_1
17. $H(c,\alpha) \leftrightarrow G(\alpha,c)$	17. IU, T_1
18. $G(\beta,c)$	18. Mod pon 15,16
19. $G(\alpha,c)$	19. Mod pon 4,17
20. $G(\alpha,c)G(\beta,c) \rightarrow \alpha = \beta$	20. IU, $A_{4.2}$
21. $\alpha = \beta$	21. Conj inf, mod pon 18,19,20
22. $(1) \rightarrow \alpha = \beta$	22. Ded prin
23. T_2	23. PGU [applied 4 times]

L_5: $(a,\alpha)\{W(a,\alpha) \rightarrow (\exists b,c)[J(b,c;\alpha)D(a,b,c)]\}$

Proof:

1. $W(a,\alpha)$	1. Assumption
2. $(\exists b,c,d)[J(b,c,d;\alpha)D(b,c,d)]$	2. IU, $A_{2.1}$

3. $J(e,f,g;\alpha)D(e,f,g)$	3. IE 2
4. $J(e,f,g;\alpha)$	4. Simp 3
5. $J(a,e,f,g;\alpha)$	5. Conj inf 1,4
6. $(5) \rightarrow \sim D(a,e,f,g)$	6. IU, $A_{2.2}$
7. $\sim D(a,e,f,g)$, i.e., $\sim[D(e,f,g)(a \neq e)(a \neq f)(a \neq g)]$	7. Mod pon 5,6
8. $D(e,f,g) \rightarrow \sim[(a \neq e)(a \neq f)(a \neq g)]$	8. Subst. VSF $[\sim(PQ) \leftrightarrow (P \rightarrow \sim Q)]$
9. $D(e,f,g)$	9. Simp 3
10. $\sim[(a \neq e)(a \neq f)(a \neq g)]$	10. Mod pon 9,8
11. $(a = e) \vee (a = f) \vee (a = g)$	11. Subst. VSF
12. $a = e$	12. Assumption
13. $D(a,f,g)$	13. IS 12,9
14. $J(f,g;\alpha)$	14. Simp 4
15. $(\exists b,c)[J(b,c;\alpha)D(a,b,c)]$	15. Conj inf 13,14 and PGE
16. $a = e \rightarrow (15)$	16. Ded prin [disch assump 12]
17. $a = f \rightarrow (15)$	17. Similarly, Steps 12 to 16
18. $a = g \rightarrow (15)$	18. Similarly, Steps 12 to 16
19. $(11) \rightarrow (15)$	19. Inf by cases
20. (15)	20. Mod pon 11,19
21. $W(a,\alpha) \rightarrow (15)$	21. Ded prin [disch assump 1]
22. L_5	22. PGU

T_3: $(a,b,\alpha,\beta)[H(a,\alpha)H(b,\beta)W(a,\beta) \rightarrow W(b,\alpha)]$

Proof:

1. $H(a,\alpha)H(b,\beta)W(a,\beta)$	1. Assumption
2. $W(a,\beta)$	2. Simp 1
3. $W(a,\beta) \rightarrow (\exists b,c)[J(b,c;\beta)D(a,b,c)]$	3. IU, L_5
4. $(\exists b,c)[J(b,c;\beta)D(a,b,c)]$	4. Mod pon 2,3
5. $J(e,f;\beta)D(a,e,f)$	5. IE 4
6. $a \neq e$	6. Simp 5
7. $H(a,\alpha)H(e,\alpha) \rightarrow a = e$	7. IU, $A_{3.2}$
8. $\sim[H(a,\alpha)H(e,\alpha)]$	8. Contrapos inf 7,6
9. $H(a,\alpha) \rightarrow \sim H(e,\alpha)$	9. Subst. VSF
10. $H(a,\alpha)$	10. Simp 1
11. $\sim H(e,\alpha)$	11. Mod pon 10,9
12. $(\exists \beta,b)\sim\sim[W(b,\alpha)J(e,b;\beta)]$	12. Quant conv 11
13. $\sim\sim[W(m,\alpha)J(e,m;\mu)]$	13. IE 12
14. $W(m,\alpha)J(e,m;\mu)$	14. Subst. VSF
15. $\sim[W(e,\alpha)J(a,e;\beta)]$	15. IU 10

16. $J(a,e;\beta) \rightarrow \sim W(e,\alpha)$	16. Subst. VSF
17. $W(e,\beta)$	17. Simp 5
18. $\sim W(e,\alpha)$	18. Conj inf 2,17 and mod pon 16
19. $W(m,\alpha)$	19. Simp 14
20. $J(e,m;\mu)$	20. Simp 14
21. $(\exists\delta)J(e,m;\delta)$	21. PGE
22. $\sim W(e,\alpha)W(m,\alpha)(\exists\delta)J(e,m;\delta)$	22. Conj inf 18,19,21
23. $(22) \rightarrow (\exists c,\gamma)[W(c,\alpha)J(e,c;\gamma)(m \neq c)]$	23. IU, $A_{5.1}$
24. $(\exists c,\gamma)[W(c,\alpha)J(e,c;\gamma)(m \neq c)]$	24. Mod pon 22,23
25. $W(n,\alpha)J(e,n;\nu)(m \neq n)$	25. IE 24
26. $\sim W(e,\alpha) \rightarrow (\exists b)[W(b,\alpha)(\beta)\sim J(e,b;\beta)]$	26. IU, $A_{5.2}$
27. $(\exists b)[W(b,\alpha)(\beta)\sim J(e,b;\beta)]$	27. Mod pon 18,26
28. $W(k,\alpha)(\beta)\sim J(e,k;\beta)$	28. IE 27
29. $\sim[W(k,\alpha)J(a,k;\phi)]$	29. IU 10
30. $W(k,\alpha) \rightarrow \sim J(a,k;\phi)$	30. Subst. VSF
31. $W(k,\alpha)$	31. Simp 28
32. $\sim J(a,k;\phi)$	32. Mod pon 31,30
33. $(\gamma)\sim J(a,k;\gamma)$	33. PGU
34. $(\beta)\sim J(e,k;\beta)$ or what is the same, $(\theta)\sim J(e,k;\theta)$	34. Simp 28 [change of dummy variable for what follows]
35. $J(a,e;\beta)(a \neq e)(\gamma)\sim J(a,k;\gamma)(\theta)\sim J(e,k;\theta)$	35. Conj inf 2,17,6,33,34
36. $(\exists b,c)[J(b,c;\beta)(b \neq c)(\gamma)\sim J(b,k;\gamma)(\theta)\sim J(c,k;\theta)]$	36. PGE
37. $(36) \rightarrow H(k,\beta)$	37. IU, L_4
38. $H(k,\beta)$	38. Mod pon 36,37
39. $H(b,\beta)$	39. Simp 1
40. $H(b,\beta)H(k,\beta) \rightarrow b = k$	40. IU, $A_{3.2}$
41. $b = k$	41. Conj inf 38,39 and mod pon 40
42. $W(b,\alpha)$	42. IS 41,31
43. $(1) \rightarrow W(b,\alpha)$	43. Ded prin
44. T_3	44. PGU [four times]

T_4: $(a,b,c,d,\alpha,\beta,\gamma,\delta)\{H(a,\alpha)H(b,\beta)H(c,\gamma)H(d,\delta) \rightarrow [J(a,b,c;\delta) \rightarrow J(d;\alpha,\beta,\gamma)]\}$

Proof:

1. $H(a,\alpha)H(b,\beta)H(c,\gamma)H(d,\delta)$	1. Assumption
2. $J(a,b,c;\delta)$	2. Assumption
3. $W(a,\delta)$	3. Simp 2

4. $H(a,\alpha)H(d,\delta)$	4. Simp 1
5. $H(a,\alpha)H(d,\delta)W(a,\delta)$	5. Conj inf 3,4
6. $(5) \to W(d,\alpha)$	6. IU, T_3
7. $W(d,\alpha)$	7. Mod pon 5,6
8. $W(d,\beta)$	8. Similarly, 3 to 7
9. $W(d,\gamma)$	9. Similarly, 3 to 7
10. $J(d;\alpha,\beta,\gamma)$	10. Conj inf 7,8,9
11. $J(a,b,c;\delta) \to J(d;\alpha,\beta,\gamma)$	11. Ded prin [disch assump 2]
12. $(1) \to (11)$	12. Ded prin [disch assump 1]
13. T_4	13. PGU [eight times]

DUALITY

Every unabbreviated statement in the geometry can be built up by means of individual variables: $x,y,z, \ldots$ and predicate constants L, P, W and the logical symbols: $\to$, $=$, $\sim$, etc. The predicate constants L and P are called *mutually dual*. We call L the dual of P, and P the dual of L. W is taken to be its own dual.

If $\mathfrak{F}$ is a statement in the geometry, then its dual, $\mathfrak{F}^*$, is formed as follows:

1. Express $\mathfrak{F}$ in unabbreviated form.
2. Replace each predicate constant by its dual.
3. Express the result in abbreviated form if desired.

For example,

$A_{2.1}$: $(\alpha)(\exists b,c,d)[J(b,c,d;\alpha)D(b,c,d)]$.

After Step 1, we have

$A_{2.1}$: $(x)(\exists y,z,w)\{L(x) \to [P(y)P(z)P(w)W(y,x)W(z,x)W(w,x)(y \neq z)(z \neq w)(w \neq y)]\}$.

After Step 2, we have

$A_{2.1}^*$: $(x)(\exists y,z,w)\{P(x) \to [L(y)L(z)L(w)W(y,x)W(z,x)W(w,x)(y \neq z)(z \neq w)(w \neq y)]\}$.

After Step 3, we have

$A_{2.1}^*$: $(a)(\exists \beta,\gamma,\delta)[J(a;\beta,\gamma,\delta)D(\beta,\gamma,\delta)]$.

For convenience of later reference, the axioms and their duals are listed:

Axiom	*Dual*
A_0: $(x,y)[W(x,y) \leftrightarrow W(y,x)]$	A_0^*: $(x,y)[W(x,y) \leftrightarrow W(y,x)]$
A_1: $(\exists x)L(x)$	A_1^*: $(\exists x)P(x)$
$A_{2.1}$: $(\alpha)(\exists b,c,d)$ $[J(b,c,d;\alpha)D(b,c,d)]$	$A_{2.1}^*$: $(a)(\exists \beta,\gamma,\delta)$ $[J(a;\beta,\gamma,\delta)D(\beta,\gamma,\delta)]$

$A_{2.2}$: $(\alpha,b,c,d,e)[J(b,c,d,e;\alpha) \rightarrow \sim D(b,c,d,e)]$

$A_{2.2}{}^*$: $(a,\beta,\gamma,\delta,\epsilon)[J(a;\beta,\gamma,\delta,\epsilon) \rightarrow \sim D(\beta,\gamma,\delta,\epsilon)]$

$A_{3.1}$: $(\alpha)(\exists a)H(a,\alpha)$

$A_{3.1}{}^*$: $(a)(\exists\alpha)G(\alpha,a)$

$A_{3.2}$: $(\alpha,a,b)[H(a,\alpha)H(b,\alpha) \rightarrow a = b]$

$A_{3.2}{}^*$: $(a,\alpha,\beta)[G(\alpha,a)G(\beta,a) \rightarrow \alpha = \beta]$

$A_{4.1}$: $(a)(\exists\alpha)G(\alpha,a)$

$A_{4.1}{}^*$: $(\alpha)(\exists a)H(a,\alpha)$

$A_{4.2}$: $(a,\alpha,\beta)[G(\alpha,a)G(\beta,a) \rightarrow \alpha = \beta]$

$A_{4.2}{}^*$: $(\alpha,a,b)[H(a,\alpha)H(b,\alpha) \rightarrow a = b]$

$A_{5.1}$: $(a,\alpha,b)\{\sim W(a,\alpha)W(b,\alpha)(\exists\beta)J(a,b;\beta) \rightarrow (\exists c,\gamma)[W(c,\alpha)J(a,c;\gamma)(b \neq c)]\}$

$A_{5.1}{}^*$: $(\alpha,a,\beta)\{\sim W(a,\alpha)W(a,\beta)(\exists b)J(b;\alpha,\beta) \rightarrow (\exists\gamma,c)[W(a,\gamma)J(c;\alpha,\gamma)(\beta \neq \gamma)]\}$

$A_{5.2}$: $(a,\alpha)\{\sim W(a,\alpha) \rightarrow (\exists b)[W(b,\alpha)(\beta)\sim J(a,b;\beta)]\}$

$A_{5.2}{}^*$: $(\alpha,a)\{\sim W(a,\alpha) \rightarrow (\exists\beta)[W(a,\beta)(b)\sim J(b;\alpha,\beta)]\}$

The next task is to prove that the dual of any axiom is a theorem.

$T_5[A_0{}^*]$:

Proof is trivial.

$T_6[A_1{}^*]$: $(\exists x)P(x)$

Proof:

1. $(\alpha)(\exists a)H(a,\alpha)$ — 1. $A_{3.1}$
   The unabbreviated form is:
   $(x)(\exists y)[L(x) \rightarrow P(y)H(x,y)]$
2. $(\exists x)L(x)$ — 2. A_1
3. $L(z)$ — 3. IE 2
4. $(\exists y)[L(z) \rightarrow P(y)H(z,y)]$ — 4. IU 1
5. $L(z) \rightarrow P(w)H(z,w)$ — 5. IE 4
6. $P(w)H(z,w)$ — 6. Mod pon 3,5
7. $P(w)$ — 7. Simp
8. $(\exists x)P(x)$ — 8. PGE

$T_7[A_{2.1}{}^*]$: $(a)(\exists\beta,\gamma,\delta)[J(a;\beta,\gamma,\delta)D(\beta,\gamma,\delta)]$

Proof:

1. $(\exists\alpha)G(\alpha,m)$ — 1. IU, $A_{4.1}$
2. $G(\mu,m)$ — 2. IE
3. $(\exists b,c,d)[J(b,c,d;\mu)D(b,c,d)]$ — 3. IU, $A_{2.1}$
4. $J(e,f,g;\mu)D(e,f,g)$ — 4. IE
5. $(\exists\alpha)G(\alpha,e)$ — 5. IU, $A_{4.1}$
6. $G(\epsilon,e)$ — 6. IE
7. $G(\epsilon,e) \leftrightarrow H(e,\epsilon)$ — 7. IU, T_1

8. $H(e,\epsilon)$	8. Mod pon 6,7
9. $H(f,\phi)$	9. Similarly, 5 to 8
10. $H(g,\gamma)$	10. Similarly, 5 to 8
11. $G(\mu,m) \leftrightarrow H(m,\mu)$	11. IU, T_1
12. $H(m,\mu)$	12. Mod pon 2,11
13. $H(e,\epsilon)H(f,\phi)H(g,\gamma)H(m,\mu)$	13. Conj inf 8,9,10,12
14. $(13) \rightarrow [J(e,f,g;\mu) \rightarrow J(m;\epsilon,\phi,\gamma)]$	14. IU, T_4
15. $J(e,f,g;\mu) \rightarrow J(m;\epsilon,\phi,\gamma)$	15. Mod pon, 13,14
16. $J(e,f,g;\mu)$	16. Simp 4
17. $J(m;\epsilon,\phi,\gamma)$	17. Mod pon 16,15
18. $\epsilon = \phi$	18. Assumption
19. $H(f,\epsilon)$	19. IS 18,9
20. $H(e,\epsilon)H(f,\epsilon) \rightarrow e = f$	20. IU, $A_{3.2}$
21. $e = f$	21. Mod pon 8,19,20
22. $\epsilon = \phi \rightarrow e = f$	22. Ded prin [disch assump 18]
23. $e \neq f$	23. Simp 4
24. $\epsilon \neq \phi$	24. Contrapos inf 22,23
25. $\epsilon \neq \gamma$	25. Similarly, 18 to 24
26. $\phi \neq \gamma$	26. Similarly, 18 to 24
27. $J(m;\epsilon,\phi,\gamma)D(\epsilon,\phi,\gamma)$	27. Conj inf 17,24,25,26
28. $(\exists\beta,\gamma,\delta)[J(m;\beta,\gamma,\delta)D(\beta,\gamma,\delta)]$	28. PGE
29. T_7	29. PGU

$T_8[A_{2.2}{}^*]$: $(a,\beta,\gamma,\delta,\epsilon)[J(a;\beta,\gamma,\delta,\epsilon) \rightarrow \sim D(\beta,\gamma,\delta,\epsilon)]$

Proof:

1. $J(a;\beta,\gamma,\delta,\epsilon)$	1. Assumption
2. $(\exists a)H(a,\beta)$	2. IU, $A_{3.1}$
3. $H(b,\beta)$	3. IE 2
4. $(\exists\alpha)G(\alpha,a)$	4. IU, $A_{4.1}$
5. $G(\alpha,a)$	5. IE 4
6. $G(\alpha,a) \leftrightarrow H(a,\alpha)$	6. IU, T_1
7. $H(a,\alpha)$	7. Mod pon 5,6
8. $W(a,\beta)$	8. Simp 1
9. $H(a,\alpha)H(b,\beta)W(a,\beta) \rightarrow W(b,\alpha)$	9. IU, T_3
10. $W(b,\alpha)$	10. Mod pon 7,3,8,9
11. $H(c,\gamma)$	11. } Similarly, 2,3
12. $H(d,\delta)$	12. } Similarly, 2,3
13. $H(e,\epsilon)$	13. } Similarly, 2,3
14. $W(c,\alpha)$	14. } Similarly, 3 to 10
15. $W(d,\alpha)$	15. } Similarly, 3 to 10
16. $W(e,\alpha)$	16. } Similarly, 3 to 10

17. $J(b,c,d,e;\alpha) \rightarrow \sim D(b,c,d,e)$	17. IU, $A_{2.2}$
18. $\sim D(b,c,d,e)$	18. Mod pon 10,14,15,16,17
19. $b = c \vee b = d \vee b = e \vee c = d \vee c = e \vee d = e$	19. Subst. VSF
20. *Case* I: $b = c$	20. Assumption
21. $H(b,\gamma)$	21. IS 20,11
22. $H(b,\gamma) \leftrightarrow G(\gamma,b)$	22. IU, T_1
23. $G(\gamma,b)$	23. Mod pon 21,22
24. $G(\beta,b)$	24. Similarly from 3
25. $G(\beta,b)G(\gamma,b) \rightarrow \beta = \gamma$	25. IU, $A_{4.2}$
26. $\beta = \gamma$	26. Mod pon 24,23,25
27. $D(\beta,\gamma,\delta,\epsilon) \rightarrow \beta \neq \gamma$	27. VSF $[AB \rightarrow A]$
28. $\sim D(\beta,\gamma,\delta,\epsilon)$	28. Contrapos inf 27,26
29. $b = c \rightarrow \sim D(\beta,\gamma,\delta,\epsilon)$	29. Ded prin [disch assump Case I]

Similarly for cases II, . . . , VI:

30. $b = d \rightarrow (28)$	30. Similarly, 20 to 29
31. $b = e \rightarrow (28)$	31. Similarly, 20 to 29
32. $c = d \rightarrow (28)$	32. Similarly, 20 to 29
33. $c = e \rightarrow (28)$	33. Similarly, 20 to 29
34. $d = e \rightarrow (28)$	34. Similarly, 20 to 29
35. $(19) \rightarrow (28)$	35. Inf by cases 29, . . . ,34
36. $\sim D(\beta,\gamma,\delta,\epsilon)$	36. Mod pon 19,35
37. $J(a;\beta,\gamma,\delta,\epsilon) \rightarrow \sim D(\beta,\gamma,\delta,\epsilon)$	37. Ded prin [disch assump 1]
38. T_8	38. PGU

NOTE: While (28) and (36) are identical formulas, it is important to understand that (28) is dependent on both of the assumptions (1) and (20), whereas (36) is dependent only on assumption (1). In an informal proof, Steps 30 to 36 would be collapsed into one brief comment.

T_9: The duals $A_{3.1}{}^*$, $A_{3.2}{}^*$, $A_{4.1}{}^*$, $A_{4.2}{}^*$ are theorems. Indeed, they are axioms.

$T_{10}[A_{5.1}{}^*]$: $(\alpha,a,\beta)\{\sim W(a,\alpha)W(a,\beta)(\exists b)J(b;\alpha,\beta) \rightarrow (\exists\gamma,c)[W(a,\gamma)J(c;\alpha,\gamma)(\beta \neq \gamma)]\}$

Proof:

1. $\sim W(a,\alpha)W(a,\beta)(\exists b)J(b;\alpha,\beta)$	1. Assumption
2. $(\exists b)J(b;\alpha,\beta)$	2. Simp 1
3. $J(d;\alpha,\beta)$	3. IE

4. $W(d,\alpha)$ — 4. Simp 3
5. $W(d,\beta)$ — 5. Simp 3
6. $W(a,\beta)$ — 6. Simp 1
7. $J(a,d;\beta)$ — 7. Conj inf 6,5
8. $(\exists\beta)J(a,d;\beta)$ — 8. PGE
9. $\sim W(a,\alpha)$ — 9. Simp 1
10. $\sim W(a,\alpha)W(d,\alpha)(\exists\beta)J(a,d;\beta)$ — 10. Conj inf 9,4,3
11. $(10) \to (\exists c,\gamma)[W(c,\alpha)J(a,c;\gamma)(d \neq c)]$ — 11. IU, $A_{5.1}$
12. $(\exists c,\gamma)[W(c,\alpha)J(a,c;\gamma)(d \neq c)]$ — 12. Mod pon 10,11
13. $W(f,\alpha)J(a,f;\phi)(d \neq f)$ — 13. IE

NOTE: "$W(f,\alpha)J(a,f;\phi)$" is "$W(a,\phi)J(f;\alpha,\phi)$" by definitions, and commutativity and associativity properties of conjunction.

14. $\phi = \beta$ — 14. Assumption
15. $W(f,\phi)$ — 15. Simp 13
16. $W(f,\beta)$ — 16. IS 14,15
17. $J(d,f;\beta)$ — 17. Conj inf 5,16
18. $W(f,\alpha)$ — 18. Simp 13
19. $J(d,f;\alpha)$ — 19. Conj inf 4,18
20. $d \neq f$ — 20. Simp 13
21. $J(d,f;\alpha)J(d,f;\beta)(d \neq f) \to \alpha = \beta$ — 21. IU, T_2
22. $\alpha = \beta$ — 22. Mod pon 19,17,20,21
23. $\alpha \neq \beta$ — 23. ISC 6,9
24. $(\alpha = \beta)(\alpha \neq \beta)$ CONTRA — 24. Conj inf
25. $\phi \neq \beta$ — 25. CONTRA 24 disch assump 14 and establishes 25
26. $W(a,\phi)J(f;\alpha,\phi)$ — 26. Simp 13 [see note]
27. $W(a,\phi)J(f;\alpha,\phi)(\beta \neq \phi)$ — 27. Conj inf 26,25
28. $(\exists\gamma,c)[W(a,\gamma)J(f;\alpha,\gamma)(\beta \neq \gamma)]$ — 28. PGE
29. $(1) \to (28)$ — 29. Ded prin
30. T_{10} — 30. PGU

Before attempting a proof of $A_{5.2}$*, it will be convenient to have the following theorem:

T_{11}: $(a,b,\alpha,\beta)[H(a,\alpha)H(b,\beta)(\delta)\sim J(a,b;\delta) \to W(a,\beta)W(b,\alpha)]$

Proof:

1. $H(a,\alpha)H(b,\beta)(\delta)\sim J(a,b;\delta)$ — 1. Assumption
2. $\sim W(a,\beta)$ — 2. Assumption
3. $(\exists\beta,\gamma,\delta)[J(a;\beta,\gamma,\delta)D(\beta,\gamma,\delta)]$ — 3. IU, $A_{2.1}$*

4. $J(a;\lambda,\mu,\nu)D(\lambda,\mu,\nu)$	4. IE
5. $W(a,\lambda)$	5. Simp 4
6. $(\delta)\sim J(a,b;\delta)$	6. Simp 1
7. $\sim J(a,b;\lambda)$	7. IU 6
8. $W(a,\lambda) \rightarrow \sim W(b,\lambda)$	8. Subst. VSF 7
9. $\sim W(b,\lambda)$	9. Mod pon 5,8
10. $a \neq b$	10. ISC 5,9
11. $H(b,\beta)H(a,\beta) \rightarrow a = b$	11. IU, $A_{2.2}$
12. $\sim[H(b,\beta)H(a,\beta)]$	12. Contrapos inf 11,10
13. $H(b,\beta) \rightarrow \sim H(a,\beta)$	13. Subst. VSF
14. $H(b,\beta)$	14. Simp 1
15. $\sim H(a,\beta)$	15. Mod pon 14,13
16. $\sim H(a,\beta) \rightarrow (\exists\phi,f)[W(f,\beta)J(a,f;\phi)]$	16. IU, L_3
17. $(\exists\phi,f)[W(f,\beta)J(a,f;\phi)]$	17. Mod pon 15,16
18. $W(f,\beta)J(a,f;\phi)$	18. IE
19. $J(a,f;\phi)$	19. Simp
20. $\sim[W(f,\beta)J(b,f;\theta)]$	20. IU 14
21. $W(f,\beta) \rightarrow \sim J(b,f;\theta)$	21. Subst. VSF
22. $W(f,\beta)$	22. Simp 18
23. $\sim J(b,f;\theta)$	23. Mod pon
24. $(\theta)\sim J(b,f;\theta)$	24. PGU
25. $a \neq f$	25. ISC 22,2
26. $J(a,f;\phi)(a \neq f)(\delta)\sim J(a,b;\delta)$ $(\theta)\sim J(b,f;\theta)$	26. Conj inf 19,25,6,24
27. $(\exists a,f)(26)$	27. PGE
28. $(27) \rightarrow H(b,\phi)$	28. IU, L_4
29. $H(b,\phi)$	29. Mod pon
30. $H(b,\phi) \leftrightarrow G(\phi,b)$	30. IU, T_1
31. $G(\phi,b)$	31. Mod pon
32. $H(b,\beta) \leftrightarrow G(\beta,b)$	32. IU, T_1
33. $G(\beta,b)$	33. Mod pon 14, 32
34. $G(\phi,b)G(\beta,b) \rightarrow \phi = \beta$	34. IU, $A_{4.2}$
35. $\phi = \beta$	35. Mod pon 31,33,34
36. $W(a,\phi)$	36. Simp 19
37. $W(a,\beta)$	37. IS 35,36
38. $W(a,\beta)\sim W(a,\beta)$ CONTRA	38. Conj inf [Assump 2 disch]
39. $W(a,\beta)$	39. By the indirect proof, 2 to 38
40. $W(b,\alpha)$	40. Similarly, 2 to 39
41. $W(a,\beta)W(b,\alpha)$	41. Conj inf
42. $(1) \rightarrow (41)$	42. Ded prin
43. T_{11}	43. PGU

$T_{12}[A_{5.2}{}^*]$: $(\alpha,a)\{\sim W(a,\alpha) \rightarrow (\exists\beta)[W(a,\beta)(b)\sim J(b;\alpha,\beta)]\}$

Proof:

1. $\sim W(a,\alpha)$	1. Assumption
2. $\sim W(a,\alpha) \rightarrow (\exists b)[W(b,\alpha)(\beta)\sim J(a,b;\beta)]$	2. IU, $A_{5.2}$
3. $(\exists b)[W(b,\alpha)(\beta)\sim J(a,b;\beta)]$	3. Mod pon
4. $W(c,\alpha)(\beta)\sim J(a,c;\beta)$	4. IE
5. $(\exists\alpha)G(\alpha,a)$	5. IU, $A_{4.1}$
6. $G(\phi,a)$	6. IE
7. $G(\xi,c)$	7. Similarly 5,6
8. $G(\phi,a) \leftrightarrow H(a,\phi)$	8. IU, T_1
9. $H(a,\phi)$	9. Mod pon
10. $H(c,\xi)$	10. Similarly 6,8,9
11. $(\beta)\sim J(a,c;\beta)$ or $(\delta)\sim J(a,c;\delta)$	11. Simp 4
12. $H(a,\phi)H(c,\xi)(\delta)\sim J(a,c;\delta) \rightarrow W(a,\xi)W(c,\phi)$	12. IU, T_{11}
13. $W(a,\xi)W(c,\phi)$	13. Mod pon 9,10,11,12
14. $\sim[W(c,\alpha)J(f;\xi,\alpha)]$	14. IU 7
15. $W(c,\alpha) \rightarrow \sim J(f;\xi,\alpha)$	15. Subst. VSF
16. $W(c,\alpha)$	16. Simp 4
17. $\sim J(f;\xi,\alpha)$	17. Mod pon
18. $(b)\sim J(b;\xi,\alpha)$	18. PGU
19. $W(a,\xi)$	19. Simp 13
20. $W(a,\xi)(b)\sim J(b;\xi,\alpha)$	20. Conj inf
21. $(\exists\beta)[W(a,\beta)(b)\sim J(b;\beta,\alpha)]$	21. PGE
22. $\sim W(a,\alpha) \rightarrow (21)$	22. Ded prin
23. T_{12}	23. PGU

The proof of T_{12} completes the program of showing that the dual of every axiom of the geometry is a theorem; hence the metageometric Duality Principle can be established in the same way as in Sec. 4.2. We shall not repeat that proof, but will use the duality principle wherever convenient in the remaining demonstrations.

TRIANGLES

Abbreviations:

$F(a,b,c:\alpha,\beta,\gamma)$ for $D(a,b,c)D(\alpha,\beta,\gamma)J(b,c;\alpha)J(a,b;\gamma)J(c,a;\beta)$

$F^*(a,b,c:\alpha,\beta,\gamma)$ for $D(\alpha,\beta,\gamma)D(a,b,c)J(a;\beta,\gamma)J(b;\gamma,\alpha)J(c;\alpha,\beta)$

Either abbreviation is read "$a,b,c,\alpha,\beta,\gamma$ form a triangle".

L_6: $(a,b,c,\alpha,\beta,\gamma)[F(a,b,c:\alpha,\beta,\gamma) \leftrightarrow F^*(a,b,c:\alpha,\beta,\gamma)]$

Proof is left as an exercise.

L_7: $(a,b,c:\alpha,\beta,\gamma)[F(a,b,c:\alpha,\beta,\gamma) \rightarrow (\mu)\sim J(a,b,c;\mu)]$

Proof: [indirect]

1. $F(a,b,c:\alpha,\beta,\gamma)(\exists\mu)J(a,b,c;\mu)$	1. Assumption
2. $(\exists\mu)J(a,b,c;\mu)$	2. Simp 1
3. $J(a,b,c;\mu)$	3. IE
4. $J(a,b;\mu)$	4. Simp 3
5. $J(a,b;\gamma)(a \neq b)$	5. Simp 1
6. $(4)(5) \rightarrow \mu = \gamma$	6. IU, T_2
7. $\mu = \gamma$	7. Mod pon 4,5,6
8. $\mu = \alpha$	8. Similarly 4 to 7
9. $\alpha = \gamma$	9. IS 7,8
10. $\alpha \neq \gamma$	10. Simp 1
11. $(\alpha = \gamma)(\alpha \neq \gamma)$ CONTRA	11. Conj inf

With this contradiction, the lemma follows easily.

The next lemma is not quite the dual of L_7, but follows easily from L_7* and L_6. We state it without proof:

L_8: $(a,b,c:\alpha,\beta,\gamma)[F(a,b,c:\alpha,\beta,\gamma) \rightarrow (m)\sim J(m;\alpha,\beta,\gamma)]$

A notational convenience:

The symbol "$\triangle_i$" will be used in formulas to stand for the string of symbols:

$$a_i,b_i,c_i,\alpha_i,\beta_i,\gamma_i$$

where i is some positive integer. Sometimes "$\triangle$" will be used to stand for the string "$a,b,c,\alpha,\beta,\gamma$". The use of subscripts merely increases the supply of symbols for individual variables. For example, "f_1", "f_2", "f_3" denote different individual variables. Use of the same letter with different subscripts usually means that the objects designated have something in common.

Using the notational convenience, one can write

$$(\triangle)F(\triangle) \quad \text{for } (a,b,c,\alpha,\beta,\gamma)F(a,b,c:\alpha,\beta,\gamma).$$

Where "$\triangle$" is a triangle, that is, where one has "$F(\triangle)$", the points "a,b,c" are called *vertices*, and the lines "α,β,γ" are called *sides.*

Abbreviations:

$C(r,\triangle)$ for $\sim[W(r,\alpha) \vee W(r,\beta) \vee W(r,\gamma)]F(\triangle)(\exists\phi_1,\phi_2,\phi_3)[D(\phi_1,\phi_2,\phi_3)$
$J(r;\phi_1,\phi_2,\phi_3)W(a,\phi_1)W(b,\phi_2)W(c,\phi_3)]$

$A(\rho,\triangle)$ for $\sim[W(a,\rho) \vee W(b,\rho) \vee W(c,\rho)]F(\triangle)(\exists f_1,f_2,f_3)[D(f_1,f_2,f_3)$
$J(f_1,f_2,f_3)W(f_1,\alpha)W(f_2,\beta)W(f_3,\gamma)]$

Read "$C(r,\triangle)$" as "$\triangle$ is perspective from the perspective center r".
Read "$A(\rho,\triangle)$" as "$\triangle$ is perspective from the perspective axis ρ".

We shall consider two triangles to be distinct if there is no way of lettering the vertices so that the vertices of the two triangles are, respectively, equal. Thus if two triangles are to be considered the same, there must be some way of lettering, or ordering the vertices so that the respective vertices are equal. Since there are six different ways of pairing the vertices of one triangle with those of another, the abbreviation for equality of triangles takes the form

$$\begin{aligned}\triangle_1 = \triangle_2 \quad \text{for } &[(a_1 = a_2)(b_1 = b_2)(c_1 = c_2)]\vee\\ &[(a_1 = b_2)(b_1 = c_2)(c_1 = a_2)]\vee\\ &[(a_1 = c_2)(b_1 = a_2)(c_1 = b_2)]\vee\\ &[(a_1 = a_2)(b_1 = c_2)(c_1 = b_2)]\vee\\ &[(a_1 = b_2)(b_1 = a_2)(c_1 = c_2)]\vee\\ &[(a_1 = c_2)(b_1 = b_2)(c_1 = a_2)].\end{aligned}$$

In addition to the notion of distinct triangles, we will wish to talk about triangles that have no common vertices or sides. Hence the abbreviations:

$$\begin{aligned}N(\triangle_1,\triangle_2) \quad \text{for } &(a_1 \neq a_2)(a_1 \neq b_2)(a_1 \neq c_2)(b_1 \neq a_2)(b_1 \neq b_2)\\ &(b_1 \neq c_2)(c_1 \neq a_2)(c_1 \neq b_2)(c_1 \neq c_2)\\ N^*(\triangle_1,\triangle_2) \quad \text{for } &(\alpha_1 \neq \alpha_2)(\alpha_1 \neq \beta_2)(\alpha_1 \neq \gamma_2)(\beta_1 \neq \alpha_2)(\beta_1 \neq \beta_2)\\ &(\beta_1 \neq \gamma_2)(\gamma_1 \neq \alpha_2)(\gamma_1 \neq \beta_2)(\gamma_1 \neq \gamma_2)\end{aligned}$$

The following lemmas are stated without proof:

$$L_9\colon (\triangle_1,\triangle_2)\{F(\triangle_1)F(\triangle_2) \to [\triangle_1 = \triangle_2 \to (a_1 = a_2)\vee(a_1 = b_2)\vee(a_1 = c_2)]\}$$

The proof is easy but tedious.

$$L_{10}\colon (\triangle_1,\triangle_2)\{F(\triangle_1)F(\triangle_2) \to [N(\triangle_1,\triangle_2) \to N^*(\triangle_1,\triangle_2)]\}$$

An indirect proof is easy.

The next theorem is the formal equivalent of Theorem 11 of the informal development Sec. 4.2. Even the informal proof given in Sec. 4.2 is quite long, and so it should come as no surprise that the formal proof is exceedingly long. Much of the length arises from the need to prove that the points and lines encountered are distinct, which in turn entails many subsidiary proofs starting with the introduction of some assumption that is later discharged. The reader may wish to use one of the devices suggested in Sec. 5.19 for keeping track of such assumptions.

T_{13}: $(r)(\exists\triangle_1, \triangle_2)[C(r,\triangle_1)C(r,\triangle_2)(\triangle_1 \neq \triangle_2)]$

Proof:

1. $(\exists\phi_1,\phi_2,\phi_3)[J(r;\phi_1,\phi_2,\phi_3)D(\phi_1,\phi_2,\phi_3)]$	1. IU, $A_{2.1}$*
2. $J(r;\phi_1,\phi_2,\phi_3)D(\phi_1,\phi_2,\phi_3)$	2. IE
3. $W(r,\phi_1)$	3. Simp 2
4. $(3) \rightarrow (\exists b,c)[J(b,c;\phi_1)D(r,b,c)]$	4. IU, L_5
5. $(\exists b,c)[J(b,c;\phi_1)D(r,b,c)]$	5. Mod pon
6. $J(a_1,a_2;\phi_1)D(r,a_1,a_2)$	6. IE
7. $W(a_1,\phi_1)$	7. Simp 6
8. $J(r,a_1;\phi_1)$	8. Conj inf 3,7
9. $W(r,\phi_2)$	9. Simp 2
10. $W(a_1,\phi_2)$	10. Assumption
11. $J(r,a_1;\phi_2)$	11. Conj inf 9,10
12. $r \neq a_1$	12. Simp 6
13. $J(r,a_1;\phi_1)J(r,a_1;\phi_2)(r \neq a_1) \rightarrow \phi_1 = \phi_2$	13. IU, T_2
14. $\phi_1 = \phi_2$	14. Mod pon 8,11,12,13
15. $\phi_1 \neq \phi_2$	15. Simp 2
16. (14)(15) CONTRA	16. Conj inf [Assump 10 disch]
17. $\sim W(a_1,\phi_2)$	17. Indirect proof, 10 to 16
18. $\sim W(a_1,\phi_3)$	18. Similarly, 10 to 17
19. $(\exists\beta)J(r,a_1;\beta)$	19. PGE 8
20. $\sim W(a_1,\phi_2)W(r,\phi_2)(\exists\beta)J(a_1,r;\beta) \rightarrow (\exists c,\gamma)[W(c,\phi_2)J(a_1,c;\gamma)(r \neq c)]$	20. IU, $A_{5.1}$
21. $(\exists c,\gamma)[W(c,\phi_2)J(a_1,c;\gamma)(r \neq c)]$	21. Mod pon 17,9,19,20
22. $W(b_1,\phi_2)J(a_1,b_1;\gamma_1)(r \neq b_1)$	22. IE
23. $W(c_1,\phi_3)J(a_1,c_1;\beta_1)(r \neq c_1)$	23. Similarly 17 to 22
24. $W(f,\phi_3)J(b_1,f;\alpha_1)(r \neq f)$	24. Similarly 17 to 22

NOTE: There is no reason yet to know that "$f = c_1$" is true. This will now be proved by a long indirect proof ending at Step 123.

25. $f \neq c_1$	25. Assumption
26. $(\exists\alpha)G(\alpha,c_1)$	26. IU, $A_{4.1}$
27. $G(\mu,c_1)$	27. IE
28. $G(\mu,c_1) \leftrightarrow H(c_1,\mu)$	28. IU, T_1
29. $H(c_1,\mu)$	29. Mod pon
30. $H(b_1,\nu)$	30. Similarly 26 to 29
31. $\sim W(b_1,\phi_3)$	31. Similarly 10 to 18
32. $(31) \rightarrow (\exists d)[W(d,\phi_3)(\psi)\sim J(b_1,d;\psi)]$	32. IU, $A_{5.2}$
33. $(\exists d)[W(d,\phi_3)(\psi)\sim J(b_1,d;\psi)]$	33. Mod pon

34. $W(d,\phi_3)(\psi)\sim J(b_1,d;\psi)$	34. IE
35. $W(r,\phi_3)$	35. Simp 2
36. $W(c_1,\phi_3)$	36. Simp 23
37. $W(f,\phi_3)$	37. Simp 24
38. $W(d,\phi_3)$	38. Simp 34
39. $J(r,c_1,f,d;\phi_3) \rightarrow \sim D(r,c_1,f,d)$	39. IU, $A_{2.2}$
40. $\sim D(r,c_1,f,d)$	40. Mod pon 35,36,37,38,39
41. $(\psi)\sim J(b_1,d;\psi)$	41. Simp 34
42. $\sim J(b_1,d;\alpha_1)$	42. IU
43. $J(b_1,f;\alpha_1)$	43. Simp 24
44. $d \neq f$	44. ISC 42,43
45. $W(b_1,\phi_2)$	45. Simp 22
46. $J(r,b_1;\phi_2)$	46. Conj inf 9,45
47. $\sim J(b_1,d;\phi_2)$ or what is the same $\sim J(d,b_1;\phi_2)$	47. IU 41
48. $d \neq r$	48. ISC 46,47
49. $r \neq c_1$	49. Simp 23
50. $r \neq f$	50. Simp 24
51. $D(r,c_1,f)(d \neq r)(d \neq f) \rightarrow d = c_1$	51. Subst. VSF for 40
52. $d = c_1$	52. Mod pon 49,50,25,48,44,51
53. $(\psi)\sim J(b_1,c_1;\psi)$	53. IS 52,41
54. $H(c_1,\mu)H(b_1,\nu)(\psi)\sim J(b_1,c_1;\psi) \rightarrow W(c_1,\nu)W(b_1,\mu)$	54. IU, T_{11}
55. $W(c_1,\nu)W(b_1,\mu)$	55. Mod pon 29,30,53,54
56. $W(b_1,\mu)$	56. Simp 55
57. $W(a_1,\gamma_1)$	57. Simp 22
58. $\phi_2 \neq \gamma_1$	58. ISC 57,17
59. $W(r,\alpha_1)$	59. Assumption
60. $W(f,\alpha_1)$	60. Simp 24
61. $J(r,f;\phi_3)$	61. Conj inf 35,37
62. $J(r,f;\alpha_1)J(r,f;\phi_3)(r \neq f) \rightarrow \alpha_1 = \phi_3$	62. IU, T_2
63. $\alpha_1 = \phi_3$	63. Mod pon 59,60,61,50,62
64. $W(b_1,\alpha_1)$	64. Simp 24
65. $W(b_1,\phi_3)$	65. IS 63,64
66. $W(b_1,\phi_3)\sim W(b_1,\phi_3)$ CONTRA	66. Conj inf 31,65 [Assump 59 disch]
67. $\sim W(r,\alpha_1)$	67. Indirect proof 59 to 66
68. $\phi_2 \neq \alpha_1$	68. ISC 9,67
69. $\alpha_1 = \gamma_1$	69. Assumption

70. $\sim W(r,\alpha_1) \to (\exists\beta)[W(r,\beta)$ $(b)\sim J(b;\alpha_1,\beta)]$	70. IU, $A_{5.2}$*
71. $(\exists\beta)[W(r,\beta)(b)\sim J(b;\alpha_1,\beta)]$	71. Mod pon 67,70
72. $W(r,\xi)(b)\sim J(b;\alpha_1,\xi)$	72. IE
73. $W(a_1,\gamma_1)$	73. Repetition of 57
74. $J(a_1;\gamma_1,\phi_1)$	74. Conj inf 7,73
75. $J(a_1;\alpha_1,\phi_1)$	75. IS 69,74
76. $(b)\sim J(b;\alpha_1,\xi)$	76. Simp 72
77. $\sim J(a_1;\alpha_1,\xi)$	77. IU
78. $\phi_1 \neq \xi$	78. ISC 75,77
79. $\phi_2 \neq \xi$	79. Similarly 73 to 78
80. $\phi_3 \neq \xi$	80. Similarly 73 to 78
81. $D(\phi_1,\phi_2,\phi_3)$	81. Simp 2
82. $D(\phi_1,\phi_2,\phi_3,\xi)$	82. Conj inf 78,79,80,81
83. $J(r;\phi_1,\phi_2,\phi_3)$	83. Simp 2
84. $W(r,\xi)$	84. Simp 72
85. $J(r;\phi_1,\phi_2,\phi_3,\xi) \to \sim D(\phi_1,\phi_2,\phi_3,\xi)$	85. IU, $A_{2.2}$*
86. $\sim D(\phi_1,\phi_2,\phi_3,\xi)$	86. Mod pon 83,84,85
87. (82)(86) CONTRA	87. Conj inf [Assump 69 disch]
88. $\alpha_1 \neq \gamma_1$	88. Indirect proof, 69 to 87
89. $W(b_1,\alpha_1)$	89. Simp 24
90. $W(b_1,\gamma_1)$	90. Simp 22
91. $J(b_1;\gamma_1,\phi_2,\alpha_1,\mu) \to \sim D(\gamma_1,\phi_2,\alpha_1,\mu)$	91. IU, $A_{2.2}$*
92. $\sim D(\gamma_1,\phi_2,\alpha_1,\mu)$	92. Mod pon 90,45,89,56,91
93. $D(\gamma_1,\phi_2,\alpha_1) \to$ $\sim[(\mu \neq \gamma_1)(\mu \neq \phi_2)(\mu \neq \alpha_1)]$	93. Subst. VSF
94. $D(\gamma_1,\phi_2,\alpha_1)$	94. Conj inf 58,88,68
95. $\sim[(\mu \neq \gamma_1)(\mu \neq \phi_2)(\mu \neq \alpha_1)$	95. Mod pon
96. $\mu = \gamma_1 \vee \mu = \phi_2 \vee \mu = \alpha_1$	96. Subst. VSF
97. *Case* I: $\mu = \gamma_1$	97. Assumption
98. $H(c_1,\gamma_1)$	98. IS 97 in 29
99. $\sim[W(a_1,\gamma_1)J(c_1,a_1;\beta_1)]$	99. IU 98
100. $W(a_1,\gamma_1) \to \sim J(c_1,a_1;\beta_1)$	100. Subst. VSF
101. $\sim J(c_1,a_1;\beta_1)$	101. Mod pon 57,100
102. $J(a_1,c_1;\beta_1)$	102. Simp 23
103. (102)(101) CONTRA	103. Conj inf [Assump Case I disch]
104. $\mu \neq \gamma_1$	104. Indirect proof, 97 to 103
105. *Case* II: $\mu = \phi_2$	105. Assumption

106. $H(c_1,\phi_2)$	106. IS 105 in 29
107. $\sim[W(r,\phi_2)J(c_1,r;\phi_3)]$	107. IU 106
108. $W(r,\phi_2) \rightarrow \sim J(c_1,r;\phi_3)$	108. Subst. VSF
109. $\sim J(c_1,r;\phi_3)$	109. Mod pon 9,108
110. $J(c_1,r;\phi_3)$	110. Conj inf 36,35
111. (110)(109) CONTRA	111. Conj inf [Assump Case II disch]
112. $\mu \neq \phi_2$	112. Indirect proof, 105 to 111
113. *Case* III: $\mu = \alpha_1$	113. Assumption
114. $H(c_1,\alpha_1)$	114. IS 113 in 29
115. $\sim[W(f,\alpha_1)J(c_1,f;\phi_3)]$	115. IU 114
116. $W(f,\alpha_1) \rightarrow \sim J(c_1,f;\phi_3)$	116. Subst. VSF
117. $W(f,\alpha_1)$	117. Simp 24
118. $\sim J(c_1,f;\phi_3)$	118. Mod pon
119. $J(c_1,f;\phi_3)$	119. Conj inf 36,37
120. (119)(118) CONTRA	120. Conj inf [Assump Case III disch]
121. $\mu \neq \alpha_1$	121. Indirect proof, 113 to 120
122. $(\mu \neq \gamma_1)(\mu \neq \phi_2)(\mu \neq \alpha_1)$	122. Conj inf 104,112,121
123. (122)(95) CONTRA	123. Conj inf [Assump 25 disch'd]
124. $f = c_1$	124. Indirect proof, 25 to 123
125. $W(b_1,\phi_2)$	125. Simp 22
126. $a_1 \neq b_1$	126. ISC 125,17
127. $a_1 \neq c_1$	127. Similarly, 125 to 126
128. $b_1 \neq c_1$	128. Similarly, 125 to 126
129. $D(a_1,b_1,c_1)$	129. Conj inf 126,127,128
130. $\alpha_1 \neq \gamma_1$	130. See note

NOTE: The indirect argument (69) to (88) established "$\alpha_1 \neq \gamma_1$" subject to the assumption (25), but in fact, the subdemonstration is independent of (25), and so will not be repeated here.

131. $\alpha_1 \neq \beta_1$	131. } Similarly
132. $\beta_1 \neq \gamma_1$	132. } Similarly
133. $D(\alpha_1,\beta_1,\gamma_1)$	133. Conj inf 130,131,132
134. $J(a_1,b_1;\gamma_1)$	134. Simp 22
135. $J(b_1,f;\alpha_1)$	135. Simp 24
136. $J(b_1,c_1;\alpha_1)$	136. IS 124 in 135
137. $J(c_1,a_1;\beta_1)$	137. Simp 23

138. $F(\triangle_1)$	138. Conj inf 129,133,134,136,137
139. $W(a_1,\beta_1)$	139. Simp 23
140. $\beta_1 \neq \phi_3$	140. ISC 139,18
141. $W(r,\beta_1)$	141. Assumption
142. $W(c_1,\beta_1)$	142. Simp 23
143. $W(c_1,\phi_3)$	143. Simp 23
144. $W(r,\phi_3)$	144. Simp 2
145. $r \neq c_1$	145. Simp 23
146. $J(r,c_1;\beta_1)J(r,c_1;\phi_3)(r \neq c_1)$	146. Conj inf 141 . . . 145
147. $(146) \rightarrow \beta_1 = \phi_3$	147. IU, T_2
148. $\beta_1 = \phi_3$	148. Mod pon
149. (148)(140) CONTRA	149. Conj inf [Assump 141 disch'd]
150. $\sim W(r,\beta_1)$	150. Indirect proof, 141 to 149
151. $\sim W(r,\alpha_1)$	151. Similarly, 141 to 150
152. $\sim W(r,\gamma_1)$	152. Similarly, 141 to 150
153. (150)(151)(152)	153. Conj inf
154. $\sim[W(r,\alpha_1) \vee W(r,\beta_1) \vee W(r,\gamma_1)]$	154. Subst. VSF
155. $W(b_1,\phi_2)$	155. Simp 22
156. $W(c_1,\phi_3)$	156. Simp 23
157. $W(a_1,\phi_1)W(b_1,\phi_2)W(c_1,\phi_3)$	157. Conj inf 7,155,156
158. $D(\phi_1,\phi_2,\phi_3)J(r;\phi_1,\phi_2,\phi_3)(157)$	158. Conj inf 2,157
159. $(\exists\phi_1,\phi_2,\phi_3)(158)$	159. PGE
160. $(154)F(\triangle_1)(159)$, that is, $C(r,\triangle_1)$	160. Conj inf 154,138,159
161. $C(r,\triangle_2)$	161. Similarly, 7 to 160
162. $a_1 \neq a_2$	162. Simp 6
163. $W(b_2,\phi_2)$	163. Simp 161
164. $a_1 \neq b_2$	164. ISC 163,17
165. $W(c_2,\phi_3)$	165. Simp 161
166. $a_1 \neq c_2$	166. ISC 165,18
167. $(a_1 \neq a_2)(a_1 \neq b_2)(a_1 \neq c_2)$	167. Conj inf 162,164,166
168. $\sim[a_1 = a_2 \vee a_1 = b_2 \vee a_1 = c_2]$	168. Subst. VSF
169. $F(\triangle_2)$	169. Simp 161
170. $F(\triangle_1)F(\triangle_2) \rightarrow [\triangle_1 = \triangle_2 \rightarrow (a_1 = a_2 \vee a_1 = b_2 \vee a_1 = c_2)]$	170. IU, L_9
171. $\triangle_1 = \triangle_2 \rightarrow (a_1 = a_2 \vee a_1 = b_2 \vee a_1 = c_2)$	171. Mod pon 138,169,170
172. $\triangle_1 \neq \triangle_2$	172. Contrapos inf 171,168
173. $C(r,\triangle_1)C(r,\triangle_2)(\triangle_1 \neq \triangle_2)$	173. Conj inf 160,161,172
174. $(\exists\triangle_1,\triangle_2)(173)$	174. PGE
175. $(r)(174)$	175. PGU

Corollary 1: $(r,\Delta_1,\Delta_2,\Delta_3)[C(r,\Delta_1)C(r,\Delta_2)C(r,\Delta_3) \rightarrow \sim D(\Delta_1,\Delta_2,\Delta_3)]$
Corollary 2: $(r,\Delta_1,\Delta_2)[C(r,\Delta_1)C(r,\Delta_2)(\Delta_1 \neq \Delta_2) \rightarrow N(\Delta_1,\Delta_2)]$
Corollary 3: $(a_1,a_2, \ldots, a_7)D(a_1,a_2, \ldots, a_7)$

The last corollary follows from $A_1{}^*$ and T_{13} and Corollary 1.

The next theorem is the Desargues theorem for the geometry. It corresponds to Theorem 12 of the informal development of Sec. 4.2.

$$T_{14}\text{: } (\Delta_1,\Delta_2)\{\Delta_1 \neq \Delta_2 \rightarrow (r)[C(r,\Delta_1)C(r,\Delta_2) \rightarrow (\exists\rho)(A(\rho,\Delta_1)A(\rho,\Delta_2)H(r,\rho))]\}$$

Proof:

1. $\Delta_1 \neq \Delta_2$	1. Assumption
2. $C(r,\Delta_1)C(r,\Delta_2)$	2. Assumption
3. $(\exists\phi_1,\phi_2,\phi_3)[D(\phi_1,\phi_2,\phi_3)J(r;\phi_1,\phi_2,\phi_3)$ $W(a_1,\phi_1)W(b_1,\phi_2)W(c_1,\phi_3)]$	3. Simp 2
4. $D(\phi_1,\phi_2,\phi_3)J(r;\phi_1,\phi_2,\phi_3)W(a_1,\phi_1)$ $W(b_1,\phi_2)W(c_1,\phi_3)$	4. IE
5. $D(\xi_1,\xi_2,\xi_3)J(r;\xi_1,\xi_2,\xi_3)W(a_2,\xi_1)$ $W(b_2,\xi_2)W(c_2,\xi_3)$	5. Similarly, 3,4
6. $W(r,\xi_1)$	6. Simp 5
7. $J(r;\phi_1,\phi_2,\phi_3)$	7. Simp 4
8. $J(r;\phi_1,\phi_2,\phi_3,\xi_1) \rightarrow \sim D(\phi_1,\phi_2,\phi_3,\xi_1)$	8. IU, $A_{2.2}{}^*$
9. $\sim D(\phi_1,\phi_2,\phi_3,\xi_1)$	9. Mod pon 6,7,8
10. $D(\phi_1,\phi_2,\phi_3) \rightarrow$ $\sim[(\xi_1 \neq \phi_1)(\xi_1 \neq \phi_2)(\xi_1 \neq \phi_3)]$	10. Subst. VSF
11. $D(\phi_1,\phi_2,\phi_3)$	11. Simp 4
12. $\sim[(\xi_1 \neq \phi_1)(\xi_1 \neq \phi_2)(\xi_1 \neq \phi_3)]$	12. Mod pon 11,10
13. $\xi_1 = \phi_1 \vee \xi_1 = \phi_2 \vee \xi_1 = \phi_3$	13. Subst. VSF

Proceeding in this way, it is easy to see that there are six cases to consider; one for each way that the lines ξ_1,ξ_2,ξ_3 can be set, respectively, equal to ϕ_1,ϕ_2,ϕ_3.

14. *Case* I: $(\xi_1 = \phi_1)(\xi_2 = \phi_2)(\xi_3 = \phi_3)$	14. Assumption
15. $D(\phi_1,\phi_2,\phi_3)J(r;\phi_1,\phi_2,\phi_3)$ $W(a_2,\phi_1)W(b_2,\phi_2)W(c_2,\phi_3)$	15. IS, Simp 14,5
16. $W(b_2,\phi_1)$	16. Assumption
17. $W(a_2,\phi_1)$	17. Simp 15
18. $J(b_2,a_2;\phi_1)$	18. Conj inf 16,17
19. $J(b_2,a_2;\gamma_2)$	19. Simp 2

20. $b_2 \neq a_2$	20. Simp 2
21. $(18)(19)(20) \rightarrow \phi_1 = \gamma_2$	21. IU, T_2
22. $\phi_1 = \gamma_2$	22. Mod pon 18,19,20,21
23. $W(r,\phi_1)$	23. Simp 7
24. $W(r,\gamma_2)$	24. IS 22,23
25. $\sim[W(r,\alpha_2) \vee W(r,\beta_2) \vee W(r,\gamma_2)]$	25. Simp 2
26. $\sim W(r,\alpha_2) \sim W(r,\beta_2) \sim W(r,\gamma_2)$	26. Subst. VSF
27. $\sim W(r,\gamma_2)$	27. Simp
28. (24)(26) CONTRA	28. Conj inf [Assump 16 disch'd]
29. $\sim W(b_2,\phi_1)$	29. Indirect proof, 16 to 28
30. $\sim W(b_2,\phi_1) \rightarrow (\exists b)[W(b,\phi_1)(\beta) \sim J(b_2,b;\beta)]$	30. IU, $A_{5.2}$
31. $(\exists b)[W(b,\phi_1)(\beta) \sim J(b_2,b;\beta)]$	31. Mod pon
32. $W(d,\phi_1)(\beta) \sim J(b_2,d;\beta)$	32. IE
33. $W(d,\phi_1)$	33. Simp 32
34. $W(a_1,\phi_1)$	34. Simp 4
35. $W(a_2,\phi_1)$	35. Simp 15
36. $W(r,\phi_1)$	36. Simp 4
37. $J(r,a_1,a_2,d;\phi_1) \rightarrow \sim D(r,a_1,a_2,d)$	37. IU, $A_{2.2}$
38. $\sim D(r,a_1,a_2,d)$	38. Mod pon 36,34,35,33,37
39. $C(r,\triangle_1)C(r,\triangle_2)(\triangle_1 \neq \triangle_2) \rightarrow N(\triangle_1,\triangle_2)$	39. IU, T_{13} Cor. 2
40. $N(\triangle_1,\triangle_2)$	40. Mod pon 2,1,39
41. $a_1 \neq a_2$	41. Simp 40
42. $\sim W(r,\beta_1)$	42. Similarly, 25 to 27, which are independent of Assump 16
43. $W(a_1,\beta_1)$	43. Simp 2
44. $r \neq a_1$	44. ISC 43, 42
45. $r \neq a_2$	45. Similarly, 42 to 44
46. $D(r,a_1,a_2) \rightarrow (d = r) \vee (d = a_1) \vee (d = a_2)$	46. Subst. from VSF for 38
47. $(d = r) \vee (d = a_1) \vee (d = a_2)$	47. Mod pon 44,45,41,46
48. $(\beta) \sim J(b_2,d;\beta)$	48. Simp 32
49. $W(b_2,\phi_2)$	49. Simp 15
50. $W(r,\phi_2)$	50. Simp 4
51. $J(r,b_2;\phi_2)$	51. Conj inf 49,50
52. $\sim J(b_2,d;\phi_2)$ or what is the same $\sim J(d,b_2;\phi_2)$	52. IU 48
53. $r \neq d$	53. ISC 51,52

54. $r \neq d \rightarrow (d = a_1) \vee (d = a_2)$	54. Subst. VSF for 47
55. $(d = a_1) \vee (d = a_2)$	55. Mod pon
56. $J(b_2,a_2;\gamma_2)$	56. Simp 2
57. $\sim J(b_2,d;\gamma_2)$	57. IU 48
58. $d \neq a_2$	58. ISC 56,57
59. $d \neq a_2 \rightarrow d = a_1$	59. Subst. VSF for 55
60. $d = a_1$	60. Mod pon
61. $(\beta) \sim J(b_2,a_1;\beta)$	61. IS 60,48
62. $\sim W(b_2,\phi_3)$	62. Similarly, 16 to 29
63. $(\gamma) \sim J(b_2,c_1;\gamma)$	63. Similarly, 29 to 61
64. $a_1 \neq c_1$	64. Simp 2
65. $J(a_1,c_1;\beta_1)$	65. Simp 2
66. $J(a_1,c_1;\beta_1)(a_1 \neq c_1)$ $(\beta) \sim J(b_2,a_1;\beta)(\gamma) \sim J(b_2,c_1;\gamma)$	66. Conj inf 65,64,61,63
67. $(\exists a_1,c_1)(66)$	67. PGE
68. $(67) \rightarrow H(b_2,\beta_1)$	68. IU, L_4
69. $H(b_2,\beta_1)$	69. Mod pon
70. $H(b_1,\beta_2)$	70. Similarly, 15 to 69
71. $(\exists \alpha)G(\alpha,r)$	71. IU, $A_{4.1}$
72. $G(\rho,r)$	72. IE
73. $G(\rho,r) \leftrightarrow H(r,\rho)$	73. IU, T_1
74. $H(r,\rho)$	74. Mod pon
75. $(\exists a)H(a,\phi_2)$	75. IU, $A_{3.1}$
76. $H(f_2,\phi_2)$	76. IE
77. $H(r,\rho)H(b_1,\beta_2)H(b_2,\beta_1)H(f_2,\phi_2)$	77. Conj inf 74,70,69,76
78. $(77) \rightarrow [J(r,b_1,b_2;\phi_2) \rightarrow J(f_2;\rho,\beta_2,\beta_1)]$	78. IU, T_4
79. $J(r,b_1,b_2;\phi_2) \rightarrow J(f_2;\rho,\beta_2,\beta_1)$	79. Mod pon
80. $J(b_1,b_2;\phi_2)$	80. Simp 2
81. $W(r,\phi_2)$	81. Simp 3 or 2
82. $J(f_2;\rho,\beta_2,\beta_1)$	82. Mod pon 81,80,79
83. $H(f_1,\phi_1)$	83.–84. Similarly, 75,76
84. $H(f_3,\phi_3)$	
85. $J(f_1;\rho,\alpha_2,\alpha_1)$	85.–86. Similarly, 15 to 82
86. $J(f_3;\rho,\gamma_2,\gamma_1)$	
87. $f_1 = f_2$	87. Assumption
88. $H(f_1,\phi_1) \leftrightarrow G(\phi_1,f_1)$	88. IU, T_1
89. $G(\phi_1,f_1)$	89. Mod pon 83,88
90. $G(\phi_2,f_2)$	90. Similarly
91. $G(\phi_2,f_1)$	91. IS 87,90
92. $G(\phi_2,f_1)G(\phi_1,f_1) \rightarrow \phi_2 = \phi_1$	92. IU, $A_{4.2}$
93. $\phi_2 = \phi_1$	93. Mod pon 91,89,92
94. $\phi_2 \neq \phi_1$	94. Simp 4

95. (93)(94) CONTRA	95. Conj inf [Assump 87 disch'd]
96. $f_1 \neq f_2$	96. Indirect proof, 87 to 95
97. $f_2 \neq f_3$	97. Similarly, 87 to 96
98. $f_1 \neq f_3$	98. Similarly, 87 to 96
99. $D(f_1,f_2,f_3)$	99. Conj inf 96,97,98
100. $\sim[W(b_2,\rho)J(r,b_2;\phi_2)]$	100. IU 74
101. $J(r,b_2;\phi_2) \rightarrow \sim W(b_2,\rho)$	101. Subst. VSF 100
102. $W(b_2,\phi_2)$	102. Simp 15
103. $W(r,\phi_2)$	103. Simp 4
104. $\sim W(b_2,\rho)$	104. Mod pon 103,102,101
105. $\sim W(a_2,\rho)$	105. } Similarly, 100 to 104
106. $\sim W(c_2,\rho)$	106. }
107. (104)(105)(106)	107. Conj inf
108. $\sim[W(a_2,\rho) \vee W(b_2,\rho) \vee W(c_2,\rho)]$	108. Subst. VSF 107
109. $W(f_2,\rho)$	109. Simp 82
110. $W(f_1,\rho)$	110. Simp 85
111. $W(f_3,\rho)$	111. Simp 86
112. $J(f_1,f_2,f_3;\rho)$	112. Conj inf 109,110,111
113. $W(f_1,\alpha_2)$	113. Simp 85
114. $W(f_2,\beta_2)$	114. Simp 82
115. $W(f_3,\gamma_2)$	115. Simp 86
116. $D(f_1,f_2,f_3)J(f_1,f_2,f_3;\rho)W(f_1,\alpha_2)W(f_2,\beta_2)W(f_3,\gamma_3)$	116. Conj inf 99,112,113,114,115
117. $(\exists f_1,f_2,f_3)(116)$	117. PGE
118. $F(\triangle_2)$	118. Simp 2
119. $A(\rho,\triangle_2)$	119. Conj inf 108,118,117
120. $A(\rho,\triangle_1)$	120. Similarly
121. $A(\rho,\triangle_1)A(\rho,\triangle_2)H(r,\rho)$	121. Conj inf 120,119,74
122. $(\exists\rho)(121)$	122. PGE

NOTE: Similarly, all the other cases mentioned after Step 13 lead to (122). Hence we have (122) as an inference by cases.

123. $(2) \rightarrow (122)$	123. Ded prin
124. $(r)[(2) \rightarrow (122)]$	124. PGU
125. $(1) \rightarrow (124)$	125. Ded prin
126. T_{14}	126. PGU

Corollary 1: $(\exists a_1,a_2, \ldots ,a_{10})D(a_1,a_2, \ldots ,a_{10})$

It follows easily from T_{13} and T_{14} that

$$(r)(\exists a_2,a_3, \ldots ,a_{10})D(r,a_2,a_3, \ldots ,a_{10}).$$

This means (see abbreviation agreements)

$$(x)[P(x) \rightarrow (\exists a_2,a_3, \ldots ,a_{10})D(x,a_2, \ldots ,a_{10})].$$

By IE from $A_1{}^*$,

$$P(y).$$

So, by IU,

$$P(y) \rightarrow (\exists a_2, \ldots ,a_{10})D(y,a_2, \ldots ,a_{10}).$$

Then, by modus ponens,

$$(\exists a_2, \ldots ,a_{10})D(y,a_2, \ldots ,a_{10}),$$

and by conjunctive inference,

$$P(y)(\exists a_2, \ldots ,a_{10})D(y,a_2, \ldots ,a_{10}).$$

Finally, by PGE,

$$(\exists y)P(y)(\exists a_2, \ldots ,a_{10})D(y,a_2, \ldots ,a_{10}),$$

or with abbreviation

$$(\exists a_1,a_2, \ldots ,a_{10})D(a_1,a_2, \ldots ,a_{10}).$$

Corollary 2: $(\exists \alpha_1,\alpha_2, \ldots ,\alpha_{10})D(\alpha_1,\alpha_2, \ldots ,\alpha_{10})$

This follows from Corollary 1 by the duality principle.

Corollary 3: $(a_1,a_2, \ldots ,a_{11})\sim D(a_1,a_2, \ldots ,a_{11})$

The proof is indirect, and an easy formalization of the informal proof given in Sec. 4.2.

INDEX